Reactive Oxygen Species in Chemistry, Biology, and Medicine

NATO ASI Series

Advanced Science Institutes Series

A series presenting the results of activities sponsored by the NATO Science Committee, which aims at the dissemination of advanced scientific and technological knowledge, with a view to strengthening links between scientific communities.

The series is published by an international board of publishers in conjunction with the NATO Scientific Affairs Division

A	**Life Sciences**	Plenum Publishing Corporation
B	**Physics**	New York and London
C	**Mathematical and Physical Sciences**	D. Reidel Publishing Company Dordrecht, Boston, and Lancaster
D	**Behavioral and Social Sciences**	Martinus Nijhoff Publishers
E	**Engineering and Materials Sciences**	The Hague, Boston, Dordrecht, and Lancaster
F	**Computer and Systems Sciences**	Springer-Verlag
G	**Ecological Sciences**	Berlin, Heidelberg, New York, London,
H	**Cell Biology**	Paris, and Tokyo

Recent Volumes in this Series

Series A: Life Sciences

Reactive Oxygen Species in Chemistry, Biology, and Medicine

Edited by
Alexandre Quintanilha

University of California, Berkeley
Berkeley, California

Plenum Press
New York and London
Published in cooperation with NATO Scientific Affairs Division

Proceedings of a NATO Advanced Series Institute on
Oxygen Radicals in Biological Systems: Recent Progress
and New Methods of Study,
held September 1-14, 1985,
in Braga, Portugal

Library of Congress Cataloging in Publication Data

NATO Advanced Study Institute on Oxygen Radicals in Biological Systems: Recent Progress and New Methods of Study (1985: Braga, Portugal)
  Reactive oxygen species in chemistry, biology, and medicine.

  (NATO ASI series. Series A, Life sciences; v. 146)
  "Proceedings of a NATO Advanced Study Institute on Oxygen Radicals in Biological Systems: Recent Progress and New Methods of Study, held September 1–14, 1985, in Braga, Portugal"—T.p. verso.
  "Published in cooperation with NATO Scientific Affairs Division."
  Includes bibliographies and index.
  1. Active oxygen—Physiological effect—Congresses. 2. Radicals (Chemistry)—Congresses. I. Quintanilha, Alexandre T. II. Title. III. Series.
QP535.O1N38   1985   574.19′214   87-38498
ISBN 0-306-42808-3

© 1988 Plenum Press, New York
A Division of Plenum Publishing Corporation
233 Spring Street, New York, N.Y. 10013

Printed in the United States of America

PREFACE

A NATO Advanced Study Institute on "Oxygen Radicals in Biological Systems: Recent Progress and New Methods of Study" was held in Braga, Portugal between September 1 and September 14, 1985, in order to consider the basic chemistry and biochemistry of activated oxygen (both radical and non-radical species) and their effect in biological systems.

This book summarizes the main lectures given at this meeting. While there is no attempt to cover all the major topics in the expanding subject of oxidative mechanisms in biology, an effort has been made to provide overviews on some key aspects of this field. The authors have attempted to convey a clear picture of both what is known and what remains unclear in their respective subjects. Not only are some of the techniques used for detecting activated oxygen species described, but also their strengths and limitations. The chemistry of many of these species is discussed and the biological and/or pathological implications are carefully reviewed. The medical and therapeutic aspects of some of the well established pathways of damage and protection are analyzed. It is our hope that the material included in this book might be useful for both researchers and teachers at the graduate level.

The success of this meeting was to a large extent due to the tireless committment of Professor Alberto Amaral and Dr. Conceição Rangel; without their outstanding efforts in dealing with all the aspects of the organization, this summer school would not have been possible.

Alexandre Quintanilha
Berkeley, California

CONTENTS

FREE RADICAL MECHANISMS OF TISSUE INJURY
AND MECHANISMS OF PROTECTION

T.F. Slater and K.H. Cheeseman

Biochemistry Department, Faculty of Science
Brunel University
Uxbridge, Middlesex, U.K.

INTRODUCTION

There has been an increasing interest during the last ten years or so in the contributions of free radical reactions to the overall metabolic perturbations that result in tissue damage and disease. For a historical overview of free radical-mediated disturbances see Slater (1971); more recent references and comments can be obtained from Slater (1984).

Free radical-mediated disturbances are currently suggested to be of importance in a widening variety of diseases and types of tissue injury as illustrated in Table 1.

Table 1. Examples of diseases that have been reported to have free radical-mediated mechanisms of importance to the development of the disease.

Toxic Liver Injury (several examples)	Some Parasitic Infections
Nutritional Liver Disease (some examples)	Some Lung Disorders
Alcoholism	Photosensitisation
Inflammation	Reperfusion Injury
Rheumatoid Arthritis	Tumour Promotion
Atherosclerosis	Carcinogenesis (some examples)

A major class of reactions (called metabolic activation) is responsible for the initial step of many types of toxic liver injury. In this initial step, the parent compound is changed through interaction with enzyme-catalysed reactions to an intermediate of higher chemical reactivity, which may initiate the complex metabolic perturbations resulting in cell injury or cell death. In a number of instances, it is known that such metabolic activations result in intermediates that are free radicals (Table 2).

Table 2. Toxic agents giving free radical intermediates by metabolic activation.

CCl_4	Quinones
Aromatic nitro-compounds	Adriamycin
Aromatic amines	
Nitrosamines	
Hydrazines	

Metabolic activation to free radical intermediates often involves the NADPH-cytochrome P-450 electron transport chain that is located in the membranes of the endoplasmic reticulum (Gibson & Skett, 1986); electron efflux can be from the flavoprotein or the P-450 sites (Figure 1).

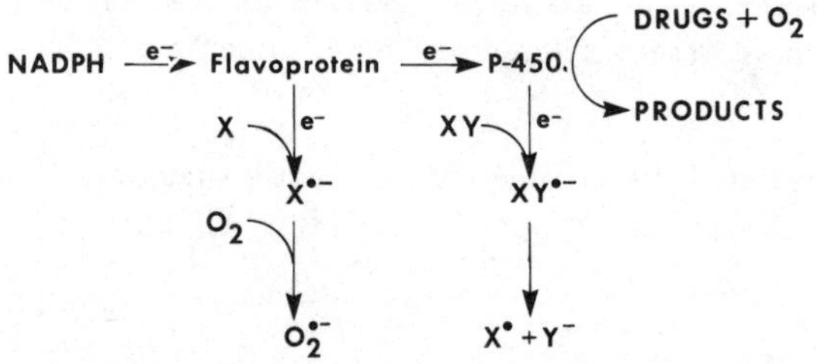

Figure 1. A much simplified illustration of a major enzymic complex that metabolically activates many substances to more toxic intermediates; some of these intermediates are free radicals. The enzymic complex requires NADPH as the reducing source, and involves a specific flavoprotein (NADPH-cytochrome P_{450} reductase; often measured by following its interaction with added cytochrome c and then called NADPH-cytochrome c reductase), and a b-type cytochrome, cytochrome P_{450}. This enzyme complex normally metabolises foreign substances to less toxic products (for short review see Gibson & Skett, 1986) by acting as a mixed function oxidase. On occasion the metabolism results in a more toxic species; the best documented example is carbon tetrachloride (see Slater, 1982). This figure shows electron efflux at both the flavoprotein and cytochrome P_{450} - loci. In the first instance, the reduction forms an intermediate that passes its electron on to oxygen; an example would be the reduction of a nitroimidazole followed by formation of superoxide anion radical. In the second instance, the reduced free radical intermediate undergoes dissociation to a new free radical species X· and the anion Y⁻; an example is the reduction of $CCl_4^-·$, followed by breakdown to $CCl_3·$ and Cl⁻; this type of reaction may produce a strongly oxidising product even though the initial reaction is a reduction.

Although the mechanism resulting in free radical formation that is illustrated in Figure 1 is reductive in the sense of involving electron transfer, the toxic free radical species produced may be a reducing *or* oxidizing species either in its primary or secondary forms (Table 3).

The discussion so far has concentrated on metabolic activation as a route for free radical formation; however, there are other mechanisms by which free radicals can be produced in cells and tissues (Table 4).

Free radicals produced in the ways outlined above can be chemically very reactive (e.g., OH·, hydroxyl), reactive (e.g., $CCl_3OO·$, trichloromethyl peroxy), rather unreactive (e.g., $O_2^-·$, superoxide) and relatively stable (e.g., DPPH·, diphenyl picryl hydrazyl).

Table 3. Formation of reducing (ii) and oxidising (iii) free radical species by metabolic activation involving electron transfer (i). In each section, an equation is written to illustrate the general features, and then followed by specific examples.

$XY + e^- \rightarrow XY^-·$ (i)

$RNO_2 + e^- \rightarrow RNO_2^-·$ (e.g., nitroimidazoles)

$O_2 + e^- \rightarrow O_2^-·$

$CCl_4 + e^- \rightarrow [CCl_4]^-·$

$XY^-· + A \rightarrow XY + A^-·$ (reduction of A)   (ii)

$RNO_2^-· + O_2 \rightarrow RNO_2 + O_2^-·$

$O_2^-· + cyt.c(Fe^{3+}) \rightarrow O_2 + cyt.c(Fe^{2+})$

$XY^-· \rightarrow X· + Y^-$ (iii)

$[CCl_4]^-· \rightarrow CCl_3· + Cl^-$

$CCl_3· + RH \rightarrow CHCl_3 + R·$ (oxidation of RH)

$O_2^-· + O_2^-· + 2H^+ \rightarrow H_2O_2 + O_2$

$H_2O_2 + Fe^{2+} + OH^- + OH·$

$OH· + RH \rightarrow H_2O + R·$ (oxidation of RH)

Table 4. Formation of free radicals in cells and tissues.

1. Radiation	
	(i) ionising
	(ii) ultra-violet
	(iii) visible (+ photosensitiser)
	(iv) thermal
2. Redox reactions	
	(i) catalysed by transitional metal ions
	(ii) catalysed by enzymes

A very reactive free radical will not be able to diffuse appreciably from its site of formation in a biological environment; free radicals such as OH· interact with biomolecules essentially under diffusion control. This is beause the OH· species is chemically so reactive that it reacts rapidly with almost anything it comes into close contact with; the 2nd-order rate constants for OH· in solution are 10^9–10^{11} $M^{-1}s^{-1}$. These high rate constants restrict the diffusion radius of OH· from its locus of formation to probably less than 2 nm. In conclusion, very reactive free radicals formed in a biological system are essentially trapped in their micro-environment: the primary disturbances produced will be within a volume of small dimensions. Thus, the cellular compartment where a reactive free radical is produced is biologically of great significance to the ensuing damage. The concept (Slater 1976) of reactivity *versus* diffusion radius is shown in Figure 2.

The discussion above on diffusion has concentrated on very reactive free radicals; it is now necessary to make the statement that the relationship between chemical reactivity and biological toxicity is not as simple as it may appear at first sight. Of course, very reactive free radicals are likely to be biologically more damaging than a relatively unreactive free radical. But a very reactive species may be less damaging to a specific biological function than a moderately reactive intermediate for the reason that the effects of the very reactive moiety are diluted by the profusion of "targets" surrounding its locus of production. A less reactive species, by being more *selective* in its chemical interactions, can be more effective in producing a particular biological response. This conclusion can be illustrated by the data in Figure 3.

The biological consequences of free radical interactions with local "targets" will depend on the cellular compartment(s) (such as the mitochondria, the nucleus, the endoplasmic reticulum, the cytosol, etc.) where the free radical is produced: on occasions (e.g., following the impact of ionizing radiation) the production of free radicals may be widely spread throughout all compartments; in other cases (e.g., a metabolic activation in the endoplasmic reticulum) the free radical may be formed largely in a relatively restricted intracellular compartment. This discussion on the compartmentation of free radical production, when taken together with the conclusions on diffusion (see Figure 2), leads to the view that the primary biological effects of free radicals formed by metabolic activation will often be closely restricted to the locus of formation of the parent free radical. Conversely, free radicals formed at random throughout the cell will not have this constraint on their immediate biological consequences.

4

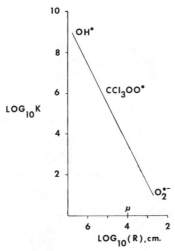

Figure 2. An illustration of how the diffusion radius of a free radical away from its locus of formation is dependent on its chemical reactivity. The values indicated should be viewed as indications of a general concept rather than precisely fixed in the numerical sense. The values have been calculated by using data for the second order rate constants of the free radicals shown with a polyunsaturated fatty acid; then, assuming that the local concentration of the free radical is extremely low compared with neighbouring biomolecules in a membrane, such as polyunsaturated fatty acids, the approximate half-life of the free radical can be calculated. In these examples, the half-lives used were 0.7×10^{-9} s (OH·), 0.7×10^{-6} s (CCl$_3$OO·) and 0.07 s (O$_2^-$·). Using these half-lives, together with a diffusion coefficient of 10^{-5} cm^2 s^{-1}, and substituting into Einstein's second diffusion equation gives the diffusion radius (R): $R^2 = 6D (t_{1/2})$ for a time equal to the half-life.

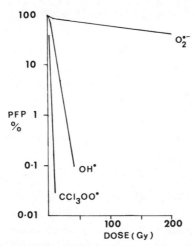

Figure 3. The inactivation of the bacteriophage T$_2$ expressed as the change in plaque-forming particles (PFP) in relation to the radiation dose (in Gray). The radiation was produced by ^{60}Co-irradiation. It can be seen that at equivalent radiation levels the effect of CCl$_3$OO· is greater than OH· and much greater than obtained with O$_2^-$·. The data are redrawn from the work of Hiller et al. (1983).

5

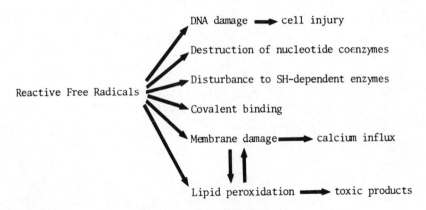

Figure 4. Major ways in which a reactive free radical can cause cell damage. This diagram should be viewed together with Figure 2 (diffusion radius).

When a suitably reactive free radical is formed in a cell, then damage can result in a number of main ways as shown in Figure 4: these pathways will be discussed briefly below with some concluding comments on mechanisms of protection.

DAMAGE TO NUCLEIC ACID

Reactive free radicals formed close to DNA (e.g., radiolysis giving a high yield of OH· in biological materials, and formed randomly throughout the sample) can attack the bases and sugar molecules of the polynucleotide. The oxidizing species OH· can abstract a hydrogen atom or add across a double bond to give a radical adduct (equations 1 and 2); the radical adduct can subsequently add O_2 (reaction 3) followed by degradative reactions.

$$\text{Base} - \text{H} + \text{OH·} \rightarrow \text{Base·} + \text{H}_2\text{O} \tag{1}$$

$$\text{Base} + \text{OH·} \rightarrow \text{Base·} - \text{OH} \tag{2}$$

$$\text{Base·} - \text{OH} + \text{O}_2 \rightarrow \text{Base} \begin{matrix} \text{OO·} \\ \text{OH} \end{matrix} \tag{3}$$

If the reactive free radical is formed in compartments of the cell other than the nucleus, then the primary interactions with DNA will be attenuated by the diffusion argument summarized in Figure 2. In consequence, a compound such as CCl_4 that is metabolically activated to a free radical intermediate mainly in the endoplasmic reticulum is not markedly mutagenic or carcinogenic in a primary sense: the intermediates CCl_3· and CCl_3OO· are too reactive to diffuse from their locus of formation to the nucleus. However, secondary and tertiary products resulting from the local attack and influence of CCl_3· may diffuse and, under certain favorable conditions, attack DNA.

If a nucleotide is positioned close to the site of formation of a reactive free radical, then it can be "attacked" in the same ways as outlined above for DNA: as a result, the biological functions of the nucleotide may be lost. An example is the destruction of

NADPH during the metabolic activation of CCl$_4$ *in vivo* (Slater et al., 1964) and *in vitro* (Slater et al., 1985).

COVALENT BINDING

Many electrophilic free radicals add across double bonds to give covalently bound adducts (see Reynolds et al., 1984; Link et al., 1984). This may be seen with the carbon-carbon centres of unsaturation in proteins, lipids, and nucleic acids. The free radical can also add directly to a heterocyclic atom or to a thiyl-radical (reactions 4 and 5); for detailed discussion of sulphur radicals see the chapter by Dr. Asmus in this book.

$$\mathrm{R\cdot} + \underset{\underset{t\text{-}bu}{|}}{\mathrm{N}} = \mathrm{O} \rightarrow \mathrm{R} - \underset{\underset{t\text{-}bu}{|}}{\mathrm{N\cdot}} - \mathrm{O} \tag{4}$$

$$\mathrm{R\cdot} + \mathrm{X} - \mathrm{S\cdot} \rightarrow \mathrm{X} - \mathrm{S} - \mathrm{R} \tag{5}$$

If the presence of the covalently bound species interferes with the normal function of the parent molecule, then perturbation of cellular structure and/or function can result. On the other hand, much of the covalent binding may be unspecific and, perhaps, in sites distant from a biologically essential molecular structure: in these circumstances the covalent binding may be relatively innocuous to the cell. Although it is easy to demonstrate that covalent binding occurs in many types of cellular injury, it has proved very difficult to establish unequivocally the quantitative significance to the injurious reaction of many of these examples of covalent binding. For example, even when covalent binding is present to only a limited extent in an injurious situation, it is hard to satisfy the criticism "a very small amount of covalent binding is damaging by selectively disturbing a few critical targets".

The identification of covalently bound adducts that may be formed *in vivo* is also another potential problem area since some adducts are relatively unstable and do not readily survive the laboratory work-up procedures.

THIOL GROUPS

Some oxidizing free radicals can react with thiol (SH) or thiol anion (S$^-$), and thereby change the ratio of SH (or S$^-$) to disulphide; this may be accompanied by serious disturbance of cellular behaviour; obvious examples are where an enzyme activity is dependent upon the presence of free thiol, or where the structure of a protein is affected by the number of disulphide bonds. For further illustration of such reactions between SH (S$^-$) with free radicals see the chapters by Dr. Willson and by Dr. Asmus in this book.

In liver the main low molecular weight (non-protein) thiol is glutathione (GSH) that is usually present in a concentration of 5-10 mM (for reviews see Kosower & Kosower, 1978; Larsson et al., 1983). The non-protein thiol fraction is known to interact with protein thiols, and changes in the protein thiol fraction are considered to be especially significant in some examples of tissue injury involving free radical intermediates (see Bellomo & Orrenius, 1985). GSH has an important role in "buffering" the cell against protein-thiol disturbances; large decreases in GSH content are often associated with an exaggeration of the cell injury.

An example where changes in protein thiols are associated with disturbances of free radical content, and with disease is in cancer of the uterine cervix (see Nöhammer et al., 1984).

LIPID PEROXIDATION

Oxidizing free radicals can abstract a hydrogen atom from an unsaturated fatty acid (which may be free or esterified as in a phospholipid) to initiate the complex sequence of reactions called lipid peroxidation. A detailed description of this degradative mechanism is given by Dr. Porter's chapter in this book, so that only a few brief comments are included here.

The hydrogen atom can be more easily abstracted from an allylic methylene-carbon than from an ethylenic or saturated carbon atom; it can be expected that the order of reactivity of a fatty acid with an oxidizing free radical will be $C_{22:6} > C_{20:4} > C_{18:3} > C_{18:2} > C_{18:1}$. Data illustrating this for the free radical $CCl_3OO\cdot$ are given in Table 5.

Table 5. The kinetics of interaction of the trichloromethylperoxy free radical ($CCl_3OO\cdot$) with the unsaturated fatty acids are illustrated here by the 2nd order rate constants (K) as determined by pulse radiolysis (Forni et al., 1983).

Fatty acid	Rate constants, K (x 10^6 $M^{-1}s^{-1}$)
Oleic	1.7
Linoleic	3.9
Linolenic	7.0
Arachidonic	7.3
Docosa Hexaenoic	8.5
9,10-deuterated-oleic	1.0

It can be seen that the K-value increases with increasing unsaturation of the fatty acid. It can also be seen that using the deuterated isomer of oleic acid results in a significant decrease in the K-value: this isotope effect suggests that hydrogen-atom abstractions is an important component of the overall interaction of $CCl_3OO\cdot$ with the unsaturated fatty acids.

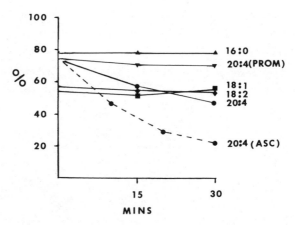

Figure 5. The changes in fatty acids of microsomal suspensions prepared from rat liver during a 30-min incubation during which time lipid peroxidation was stimulated by CCl_4. The individual fatty acid contents are expressed relative to the content of $C_{16:0}$ as 100%; the latter fatty acid does not undergo significant change in content/mg protein during these incubation conditions. Also indicated is the protective effect of the phenothiazine drug 'Promethazine' (final concentration 10 μM in preventing the loss of $C_{20:4}$. Also shown (dashed line) is the decrease in $C_{20:4}$ after incubating liver microsomes with ascorbate-ADP-iron. The data are from Ahmed & Slater (1981).

When lipid peroxidation is stimulated in samples of biological material, then the major peroxidation can be expected to involve $C_{22:6}$ and $C_{20:4}$. An example of this behaviour is shown in Figure 5 for the damaging action of CCl_4 on liver microsomal membranes.

Lipid peroxidation of a membrane such as the endoplasmic reticulum *in vivo* (or microsomes, *in vitro*) not only damages the membrane structure and function by degrading the highly unsaturated fatty acids, and thereby affecting protein:lipid interactions, but by forming breakdown products that can result in other types of membrane damage and disturbances elsewhere. Table 6 lists some of the major products of lipid peroxidation: many of these have powerful biological effects. Lipid hydroperoxides, for example, have been shown to stimulate cyclo-oxygenase when added in very low concentrations (Hemler et al., 1979); some epoxy-derivatives of arachidonate influence hormone release (Capdevila et al., 1982); the class of 4-hydroxy-alkenals which are formed along with many other carbonyl compounds during lipid peroxidation (Esterbauer et al., 1982), have a number of biological effects (Table 7).

Since products such as those described above have rather long half-lives in biological environments compared to the initiating free radical, they can diffuse from the original site of free radical-mediated injury and produce disturbances at a distance (Figure 6).

Bifunctional substances like dialdehydes and 4-hydroxy-alkenals can interact with membrance components to give cross-linking, and this can have marked effects on membrane fluidity (Dobrotsov et al., 1977; Slater, 1979).

Table 6. Products of lipid peroxidation.

Leukotrienes	Alkanals
Lipid hydroperoxides	Alkenals
Hydroxy fatty acids	4-hydroxy-Alkenals
Epoxy-fatty acids	Ketones
	Alkanes

Table 7. Biological effects of hydroxy-alkenals.

(A)	Inhibition of DNA-synthesis
(B)	Inhibition of glucose-6-phosphatase
(C)	Inhibition of adenyl cyclase
(D)	Reaction with thiols such as glutathione
(E)	Increase in capillary permeability
(F)	Inhibition of platelet aggregation by ADP
(G)	Reaction with polyamines
(H)	Stimulation of chemotaxis

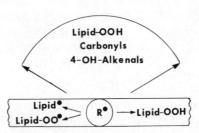

Figure 6. An illustration of the diffusion of a very reactive free radical species R· formed in a biomembrane: R· is essentially trapped in its micro-environment by virtue of its chemical reactivity. Interactions with neighbouring biomolecules fatty acids, for example, can give secondary and longer-lived intermediates that can diffuse in the plane of the membrane. Further degradation can give other products with sufficiently long half-lives to allow diffusion to other parts of the cell and even to extracellular regions. The diagram is from Slater (1986) and is a modification of that first shown in Slater (1976).

The complex family of aldehydic products of lipid peroxidation have been separated and identified by HPLC analysis of their hydrazone derivatives with dinitrophenyl hydrazine (Esterbauer, 1981). Malonaldehyde does not form a hydrazone with dinitrophenyl hydrazine, but it can be detected and measured by direct HPLC procedures (Esterbauer & Slater, 1981). Although some reports suggest that malonaldehyde is largely an artifact of the commonly used thiobarbituric acid procedure, in some biological systems it is a genuine product of lipid peroxidation.

Lipid peroxidation is initiated by hydrogen abstraction, a process requiring a suitably reactive free radical such as OH·, or peroxy- and alkoxy-radicals (ROO· and RO·). Other free radicals such as O_2^-· probably stimulate lipid peroxidation only after conversion to more reactive entities, usually by transitional metal catalysis (see Halliwell & Gutteridge, 1984). Lipid hydroperoxides can also undergo metal-catalysed degradation to reactive free radical products.

The role of Fe^{2+} in such reactions suggests that reactive free radicals can be formed in sites different to that where the parent free radical was produced (for example, O_2^-· being formed in one compartment and undergoing Fe-catalysed degradation to OH· in another compartment). As another example, lipid hydroperoxide may be formed in the endoplasmic reticulum by attack of the reactive species OH· on unsaturated fatty acid, and *may* diffuse to another site where Fe-catalysed degradation to alkoxy-radicals can occur. Of course, this latter suggestion will be affected by the influence of the enzyme GSH-peroxidase that metabolizes lipid hydroperoxides rather efficiently to the corresponding alcohols.

A corollary of the above argument is that if the normal strict compartmentation of iron in the cell is disturbed (by chemical injury or by homogenization) then Fe^{2+} can appear in "unusual" sites and may then catalyse damaging free radical reactions as outlined above. This has implications to a very important question that is often very difficult to answer: if a particular toxic treatment is associated with a stimulation of lipid peroxidation, is this a significant *cause* of injury or is it a *consequence* of injury?

MECHANISMS OF PROTECTION

A free radical scavenger is a substance that can react with a free radical to effectively prevent it from participating in the reactions it normally facilitates. If these "normal" reactions involve oxygen then the scavenger can also be called an antioxidant. The latter term refers to a substance that interferes with oxidative processes mediated by free radical intermediates. Antioxidants are conveniently classified into preventative or chain-breaking categories (Burton et al., 1983). It is obvious that some antioxidants are free radical scavengers and *vice versa*.

If the free radical that produces cell injury is highly reactive and formed in a specific cellular location (e.g., by metabolic activation), then the protective scavenger (or antioxidant) must penetrate to that precise site since the diffusion of the primary radical is restricted (Figure 2). Moreover, since the life-time of the reactive free radical is short, the scavenger must be available at the right time to compete with neighboring

Table 8. Criteria for effective scavenging of free radicals.

---

(1) The scavenger must penetrate to the relevant intracellular site where the free radical is produced.

(2) The scavenger must be in the right site at the right time since the life-times of reactive free radicals are short.

(3) The scavenger-free radical interaction must have a high second order rate constant to permit an efficient competition with biomolecules.

(4) The scavenger must attain a suitable concentration in the right site at the right time in order to compete effectively with neighbouring biomolecules.

---

biomolecules. In addition, this competition has to be efficient so that the major reaction that occurs is with the scavenger: this demands that the scavenger concentration must be locally high. This, in turn, requires that the scavenger and its reaction product have acceptable toxicity in relation to the biological system under study. These all form a demanding set of criteria (see Table 8). For a more extensive survey of these points and on biological antioxidants see Slater (1981) and Burton et al. (1985).

Acknowledgement. Much of our work summarized here has been carried out with generous financial support from the Cancer Research Campaign, the Association for International Cancer Research, and the National Foundation for Cancer Research.

REFERENCES

Ahmed, S.M., and Slater, T.F., 1981, Lipid peroxidation in microsomal fractions obtained from some rat and mouse tumours, in "Recent Advances in Lipid Peroxidation and Tissue Injury", T.F. Slater and A. Garner, eds., pp. 177-194, Brunel University, biochemistry Department, Uxbridge, U.K.

Asmus, K.D., pages 37-54 in this book.

Bellomo, G., and Orrenius, S., 1985, Altered thiol and calcium homostasis in oxidative hepatocellular injury, *Hepatology*, 5:876.

Burton, G.W., Cheeseman, K.H., Doba, T., Ingold, K.H., and Slater, T.F., 1983, Vitamin E as an antioxidant *in vitro* and *in vivo*, in "Biology of Vitamin E", R. Porter, ed., pp. 4-18, Ciba Foundation Symposium 101, Pitman Books, London.

Burton, G.W., Foster, D.O., Perly, B., Slater, T.F., Smith, I.C.P., and Ingold, K.U., 1985, Biological antioxidants, *Phil. Trans. R. Soc. Lond.*, B311:565.

Capdevila, J., Chacos, N., Falck, J.R., Manna, S., Negro-Vilar, A., and Ojeda, S.R., 1983, Novel hypothalamic arachidonate products stimulate somatostatin release from the median eminence, *Endocrinology*, 113:421.

Dobretsov, G.E., Borschevskaya, T.A., Petrov, V.A., and Vladimirov, Y.U., 1977, The increase of phospholipid bilayer rigidity after lipid peroxidation, *FEBS Letters*, 84:125.

Esterbauer, H., 1982, Aldehydic products of lipid peroxidation, in "Free Radicals, Lipid Peroxidation and Cancer", D.C.H. McBrien and T.F. Slater, eds. pp.101-122, Academic Press, London.

Esterbauer, H., and Slater, T.F., 1981, The quantitative estimation by high performance liquid chromatography of free malonaldehyde produced by peroxidising microsomes, *ICRS Medical Science*, 9:749.

Esterbauer, H., Cheeseman, K.H., Dianzani, M.U., Poli, G., and Slater, T.F., 1982, Separation and characterization of the aldehydic products of lipid peroxidation stimulated by ADP–Fe^{2+} in rat liver microsomes, *Biochem. J.*, 208:129.

Forni, L.G., Packer, J.E., Slater, T.F., and Willson, R.L., 1983, Reaction of the trichloromethyl and halothane derived peroxy radicals with unsaturated fatty acids: a pulse radiolysis study, *Chem. Biol. Interactions*, 45:171.

Gibson, G.G., and Skett, P., 1986, "Introduction to Drug Metabolism", pp. 1-292, Chapman and Hall, London.

Halliwell, B., and Gutteridge, J.M.C., 1984, Oxygen toxicity, oxygen radicals, transitional metals and disease, *Biochem. J.*, 219:1.

Hemler, M.E., Cook, H.W., and Lands, W.E.M., 1979, Prostglandin synthesis can be triggered by lipid peroxides, *Archs. Biochem. Biophys.*, 193:340.

Hiller, K-O., Hodd, P.L., and Willson, R.L., 1983, Anti-inflammatory drugs: protection of a bacterial virus as an *in vitro* biological measure of free radical activity, *Chem. Biol. Interactions*, 47:293.

Kisower, N.S., and Kosower, E.M., 1978, The glutathione stats of Cells, *Int. Rev. Cytology*, 54:109.

"Functions of Glutathione: Biochemical, Physiological, Toxicological and Clinical Aspects", A. Larsson, S. Orrenius, A. Holmgren, and B. Mannervik, eds., 1983, Raven Press, New York.

Link, B., Durk, Thiel, D., and Frank, H., 1984, Binding of trichloromethyl radicals to lipids of the hepatic endoplasmic reticulum during tetrachloromethane metabolism, *Biochem. J.*, 223:577.

Nöhammer, G., Bajardi, F., Benedetto, C., Schauenstein, E., and Slater T.F., 1984, Studies on the relationship between epithelium and stroma in sections of human uterine cervix in different pathological conditions, in "Cancer of the Uterine Cervix. Biochemical and Clinical Aspects", D.C.H. McBrien and T.F. Slater, eds., pp. 205-224, Academic Press, London.

Porter, N., pages 55-79 in this book.

Reynolds, E.S., Treinen, R.J., Farrish, H.H., and Moslen, M.T., 1984, Metabolism of ^{14}C-carbon tetrachloride to exhaled, excreted and bound metabolites, *Biochem. Pharmacol.*, 33:3363.

Slater, T.F., 1972, "Free Radical Mechanisms in Tissue Injury", pp. 1-283, Pion, London.

Slater, T.F., 1976, Biochemical pathology in microtime, in "Recent Advances in Biochemical Pathology: Toxic Liver Injury", M.U. Dianzani, G. Ugazio, and L.M. Sena, eds., pp. 381-390, Minerva Medica, Turin.

Slater, T.F., 1979, Biochemical studies of transient intermediates in relation to chemical carcinogenesis, in "Submolecular Biology and Cancer", G.E.W. Wolstenholme, D. Fitzsimons, and J. Whelan, eds., Ciba Foundation Symposium No. 67, pp. 301-328, Excerpta Medica, Amsterdam.

Slater, T.F., 1981, Free radical scavengers, in "International Workshop on (+)-cyanidanol-3 in Diseases of the Liver", H.O. Conn, ed., pp. 11-15, Royal Society of Medicine International Congress and Symposium Series No. 47, Royal Society of Medicine, London.

Slater, T.F., 1982, Activation of carbon tetrachloride: chemical principles and biological significance, in "Free Radicals, Lipid Peroxidation and Cancer", D.C.H. McBrien and T.F. Slater, eds., pp. 243-270, Academic Press, London.

Slater, T.F., 1984, Free Radical Mechanisms in Tissue Injury, *Biochem. J.*, 222:1.

Slater, T.F., Cheeseman, K.H., and Ingold, K.U., 1985, Carbon tetrachloride toxicity as a model for studying free radical mediated liver injury, *Phil. Trans. R. Soc. Lond.*, B311:633.

Slater, T.F., 1986, "Free Radical Mechanisms in Tissue Injury (2nd Edition).

Slater, T.F., Sträuli, U., and Sawyer, B., 1964, Changes in liver nucleotide concentrations in experimental liver injury. I. Carbon tetrachloride poisoning, *Biochem. J.*, 93:260.

Willson, R.L., pages 81-84 in this book.

GLUTATHIONE PEROXIDASE, SELENIUM AND VITAMIN E IN DEFENSE AGAINST REACTIVE OXYGEN SPECIES

Baldur Símonarson

Biochemistry Laboratory
Faculty of Medicine
University of Iceland, Keldur
Ármúli 30
IS-108 Reykjavík, Iceland

PART I. GLUTATHIONE PEROXIDASE, SELENIUM AND VITAMIN E IN ANIMAL NUTRITION AND VETERINARY MEDICINE*

The Veterinary and Agricultural Contribution

In Mozart's Magic Flute, the audience is misled as to the identity of heros and villains. At first, one is meant to believe that the Queen of the Night is a "goody", whose daughter Pamina is held in the evil clutches of the "baddy" Sarastro. It turns out that Sarastro is a wise and kind man, while the Queen of the Night represents the powers of darkness and superstition. For much of this century, selenium was regarded only as a toxic element, and it comes as a surprise not only to the layman, but even to biochemists and physicians, that oxygen is potentially toxic. The reversal of the roles of selenium and oxygen is, however, not as dramatic as is that of the operatic characters.

Much of our present knowledge of the roles of vitamin E and selenium comes from studies of practical problems in agriculture and veterinary medicine. Many of these problems arise from the intensive methods now used in farming. Deficiencies and excesses of vitamins and trace elements observed in veterinary medicine have contributed vastly to fundamental nutritional knowledge. It can also be seen that there is a two-way traffic in science; not only will fundamental research have later applications, but the practical knowledge gained from applied science can also help to put the basic facts into perspective. In this review, selenium and vitamin E will first be discussed in

---

*Co-authored with Gudný Eiríksdóttir and Thorsteinn Thorsteinsson, Institute for Experimental Pathology.

relation to veterinary medicine, then in relation to human health, and, finally, some fundamental properties of glutathione peroxidase will be dealt with. Vitamin E will not be discussed in as much detail as selenium. The reader is referred to Ciba Foundation Symposium 101, Biology of Vitamin E.

Veterinary and agricultural chemistry have not always been held in high esteem by the scientific establishment. J.H. van't Hoff, who was awarded the first Nobel prize in chemistry and proposed the theory of asymmetric carbon, was a lecturer at the Veterinary School in Utrecht before being appointed to the chair of chemistry at Amsterdam University in 1878. In his inaugural address, van't Hoff recalls the vicious attacks on his views made by Hermann Kolbe, one of Germany's most distinguished organic chemists:

"A Dr. J.H. van't Hoff, who is employed at the Veterinary School in Utrecht, appears to find exact chemical research not suiting his taste. He deems it more convenient to mount Pegasus (evidently loaned from the Veterinary School) and to proclaim in his 'La Chimie dans l'espace' how to him on the chemical Parnassus which he ascended in his daring flight the atoms appeared to be arranged in the Universe."

"The prosaic chemical world found little taste in these hallucinations; therefore, Dr. F. Herrmann, assistant at the Agricultural Institute in Heidelberg, attempted to achieve a wider distribution of these delusions by preparing their German edition." . . .

"It is characteristic of today's uncritical and criticism-hating time, that two virtually unknown chemists, the one from a veterinary school, the other from an agricultural institute, judge the most profound problems of chemistry which will probably never be answered. They judge these most important problems, expecially the question as to the *spatial orientation* of the atoms, with a cocksureness and insolence which can only astound a true student of natural science."

Van't Hoff's lecture is entitled "Imagination in Science", and became widely available only a few years ago (van't Hoff, 1967). His message is still valid today. Surprising results and false trails in the biochemistry of vitamin E and selenium have indeed only been understood with the application of a bit of imagination to the hard experimental facts.

Chemical Properties

Selenium (Se) belongs to the group VI of elements and is placed between sulfur and tellurium in the periodic table. It has both metallic and non-metallic properties and shows variable valence states. It resembles sulfur in many respects, but selenium is much more nucleophilic. Table 1 lists the valence states, formulae and names of some important selenium and sulfur compounds.

Selenium analysis is difficult at low levels. This has contributed to it being recognized so late as an essential element. It is also difficult to produce diets low in selenium, except by feeding grain from very low Se-areas. In recent years, much progress has been made in atomic absorption spectroscopy. Direct determination of selenium in biological material, like blood plasma, is possible in a graphite furnace. Iron in erythrocytes, liver and other tissues interferes with this method. This can be overcome by the addition of nickel as a matrix modifier and determination in a graphite furnace equipped with a

Zeeman effect corrector where a magnetic field is applied to the spectrum. Such sophisticated equipment is expensive. Great sensitivity can be obtained by use of a hydride generating system (Piwonka et al., 1985). The fluorimetric method of Watkinson (1966), where diaminonaphthalene is used to produce a fluorescent piazoselenol, still remains the workhorse of selenium analysis in biological material. Although laborious, it does not require expensive equipment. It gives adequate sensitivity, good reproducibility and freedom from interference.

Table 1

Valence States	Selenium		Sulfur	
-2	Se^{2-}	selenide	S^{2-}	sulfide
+4	SeO_3^{2-}	selenite	SO_3^{2-}	sulfite
	H_2SeO_3	selenious acid	H_2SO_3	sulfurous acid
+6	SeO_4^{2-}	selenate	SO_4^{2-}	sulfate
	H_2SeO_4	selenic acid	H_2SO_4	sulfuric acid

Toxicity and Essentiality

Selenium was first recognized as a toxic element in the mid-1930's, when it was identified as the cause of diseases of livestock which farmers called blind staggers or alkali disease, and occurred in certain areas of the United States, like South Dakota. Selenium goes through a soil-plant-animal cycle, and its availability to plants depends not on its soil concentration alone, but also on soil pH and the oxidation state of selenium. In oceanic climates, sea and rain can be a major source of selenium, and its distribution then follows that of iodine and bromine more closely than that of sulfur (Låg and Steinnes, 1978). This is an important point which is often overlooked. Certain plants are capable of concentrating Se, and when animals feed on these they succumb to selenium poisoning (Fishbein, 1983; Reddy and Massaro, 1983; Underwood, 1977).

Selenium was later shown to be an essential nutrient by Schwarz and Foltz (1957) who identified it as a factor necessary for preventing dietary necrotic degeneration of the liver in the rat. Soon, selenium responsive diseases were described for livestock: white muscle disease of sheep and cattle, hepatosis dietetica of pigs and exudative diathesis of chickens. The regional distribution of selenium in forages and grains in the United States is shown in Figure 1.

Biological Action

The functions of selenium and vitamin E were shown to be related, but it took a long time to clarify the relationship (Hoekstra, 1975). The observations which led to the idea of a role for selenium in the reaction catalyzed by glutathione peroxidase, an enzyme discovered in erythrocytes in 1957, are described by Hoekstra (1974).

Glutathione peroxidase catalyzes the reactions: $H_2O_2 + 2GSH \rightarrow 2H_2O + GSSG$ and $ROOH + 2GSH \rightarrow ROH + H_2O + GSSG$ (GSH represents glutathione, GSSG represents glutathione disulfide and ROOH represents an organic hydroperoxide). Its importance in preventing oxidative damage in the erythrocyte was immediately obvious. It was shown that the erythrocytes of selenium-deficient rats had little protection against peroxide-induced membrane damage and oxidative damage to hemoglobin. The erythrocytes of such rats had virtually no glutathione peroxidase activity. Vitamin E proved ineffective in protecting against oxidative damage in the erythrocytes of the selenium-deficient rats. This showed clearly that the biological effects of selenium and vitamin E were distinct. These observations brought down what one might call "the belt and braces hypothesis", which postulated that only one of these dietary constituents was necessary, and that one could replace the other. One could indeed extend this hypothesis by pointing out that in the well fed state, niether is essential for support. The work of Hoekstra and his colleagues culminated in the demonstration that isotopically labelled Se was extensively incorporated into the purified enzyme (Rotruck et al., 1973).

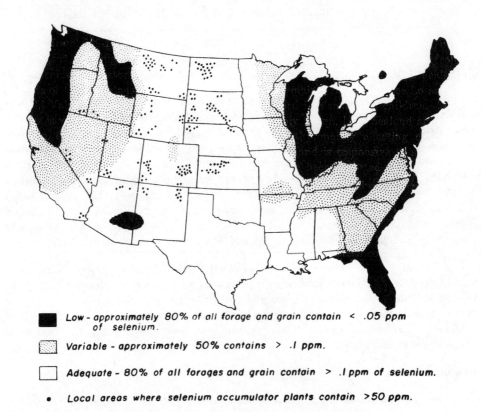

■ Low - approximately 80% of all forage and grain contain < .05 ppm of selenium.

▦ Variable - approximately 50% contains > .I ppm.

□ Adequate - 80% of all forages and grain contain > .I ppm of selenium.

• Local areas where selenium accumulator plants contain >50 ppm.

Figure 1. Regional distribution of selenium in the United States of forages and grains which are low, variable, adequate, or toxic in selenium (Ammerman & Miller, 1975).

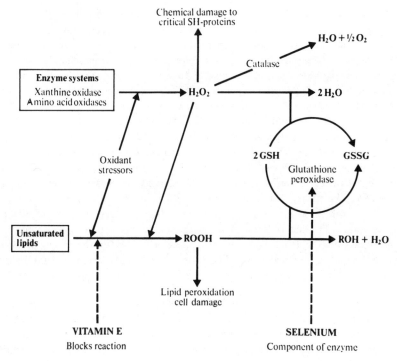

Figure 2. Postulated functions of selenium and vitamin E (Anderson & Patterson, 1976; after Hoekstra, 1975).

The interrelationship of selenium, glutathione peroxidase and vitamin E could now be explained (Hoekstra, 1975). Both the enzyme and the vitamin take part in defense against oxidative damage. Vitamin E prevents it by acting as a chain terminator in lipid peroxidation and glutathione peroxidase acts on a number of organic peroxides as well as hydrogen peroxide. The functions of selenium and vitamin E, according to Hoekstra (1975) are shown in Figure 2. Superoxide dismutase and catalase also take part in defense against reactive oxygen species, and one could as well add ascorbate, β-carotene and urate to the list.

Veterinary Aspects

Measurements of erythrocyte or whole blood glutathione peroxidase activity in farm animals give a reliable indication of their selenium status. The activity of glutathione peroxidase in the erythrocytes indicates the medium-term selenium status of an animal. It is obviously related to erythrocyte life-span, which varies from species to species, but is commonly 3 to 4 months. There is hardly any glutathione peroxidase activity in plasma. Selenium in plasma seems mainly bound to proteins. The selenium level there indicates the short-term selenium status of the animal in question, since plasma proteins have fairly short biological half-lives.

19

The dietary requirements of most species are of the order of 100 parts per billion (100 ng/g). Selenium deficiency of livestock has caused great damage in many parts of the world: Australia, New Zealand, Britain, Scandinavia, and Northwest and Northeast United States. Although high levels are found in soils in Ireland, cultivated soil tends to be acidic, making less selenium available to plants. Hay samples had an average of 6 ng/g, ranging from 8-400 ng/g. White muscle disease occurs on every tenth farm in Iceland, but the incidence varies from year to year. A trial at the Hvanneyri Experimental Farm showed that glutathione peroxidase activity in the blood of ewes fell markedly in the last trimester of pregnancy (Eiriksdóttir et al., 1981). Sheep in Iceland are fed largely on hay but are often given feed supplements and fish meal which is an excellent source of selenium. However, fish meal is rich in unsaturated lipids. The meal used for animal feeds has often suffered heat damage during processing, where lipid peroxidation could easily take place. Feeding of such meal in large amounts could well result in oxidative stress, which would be better prevented by adequate amounts of vitamin E rather than selenium. The use of such meal could easily cause nutritional problems in animals like mink, which are fed largely on fish meal. Figures for the biological availability of selenium from fish vary widely. The most likely explanation is that other trace elements, mercury and perhaps cadmium, interfere with its absorption and utilization.

An interesting and unexpected observation has been reported by Anderson et al. (1983) who found that monensin added to the feed improves the selenium status of pregnant ewes as well as the lambs born to them. Their results for lambs are shown in Figure 3. Monensin is a carboxylic polyether ionophore, widely used as an additive to feed for beef cattle. It interferes with the ruminal flora and improves feed conversion (Bergen and Bates, 1984). It is also used for controlling coccidiosis, a parasitic disease of ruminants and chicken. Monensin is toxic in high doses. It would be worthwhile studying this observation further to provide an explanation closer to the molecular level.

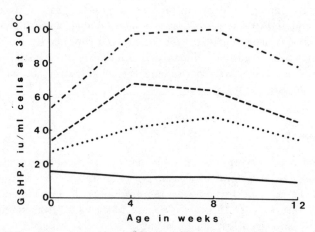

Figure 3. Blood glutathione peroxidase of lambs. The number of lambs in the control group varied between five and eight. All other groups contained at least eleven lambs. Group 1 (control) [—]; Group 2 (monensin) [⋯]; Group 3 (selenium) [---]; Group 4 (monensin and selenium) [-⋅-] (Anderson et al., 1983).

The close relation of the roles of selenium and vitamin E means that they are usually administered together in veterinary practice. Methods of prevention in low selenium areas include injections of selenium salts, slow release pellets introduced into the rumen, subcutanaeous injection of slowly dissolving barium selenate, application of Se to the soil or its inclusion in fertilizers, as well as the addition of Se salts to animal feeds. It is of interest that the veterinary profession has also been quick to realize the use of another antioxidant, preparations of bovine superoxide dismutase for the treatment of inflammations. Early studies on selenium toxicity suggested that it was carcinogenic. In the United States, the so-called Delaney Amendment prohibits the addition to foods and animal feedstuffs of any compounds shown to be carcinogenic. Since 1974, the U.S. Food and Drug Administration has gradually allowed the addition of selenium to the feed of animals. Finland has allowed the addition of selenium to animal feeds since 1969, and it is now permitted in fertilizers there. Norway, Denmark and Sweden have followed suit, but the addition of selenium to animal feedstuffs is not permitted in Iceland.

Se is present in high levels in many parts of the world, the Dakotas and Nebraska in the United States, parts of Ireland, Israel, Australia, and South America. Although the difference between a toxic dose of selenium and a therapeutic one is fairly large, care is obviously needed in its use in agriculture and veterinary medicine. Concern has recently been expressed over high Se levels in Californian soils as the result of irrigation schemes (Marshall, 1985).

It is possible to induce experimentally nutritional myopathy in calves deficient in Se and vitamin E. Polyunsaturated fatty acids which escape reduction in the rumen are thought to provide the oxidative challenge (McMurray et al., 1983). Such model studies will certainly help to elucidate the etiology of naturally occurring white muscle disease. Many farm animals, like sheep, cattle and pigs, lack an efficient mechanism for transporting vitamin E in the blood, which may be one of the explanations why they are more susceptible to vitamin E deficiency. As a result, serum levels in these animals are 10 to 100 times lower than in humans, and very sensitive analytical techniques are needed (McMurry and Blanchflower, 1979).

PART 2. GLUTATHIONE PEROXIDASE AND SELENIUM IN HUMAN HEALTH AND DISEASE*

Introduction

The identification of selenium as the toxic ingredient of certain plants poisonous to animals prompted studies of selenium levels in foods from the seleniferous areas in the United States and Canada. Much of the early interest in selenium in human nutrition was naturally directed towards its toxic effects (Reviewed by Lo and Sandi, 1980). Studies of blood Se levels in humans from various parts of the United States have shown a good correlation with the geographical variation in the selenium levels of crops as well as the selenium levels in the blood of domestic animals.

---

*Co-authored with Gudný Eiriksdóttir, Institute for Experimental Pathology.

Selenium has several industrial uses and enters our daily life more often than we realize. It is used in the glass industry, in the manufacturing of photo-electric cells, for duplicating machines and for making pigments (Fishbein, 1983). When you stop at red traffic lights, walk through an automatically opening door or take a photocopy of a recent article rather than reading it, you are encountering selenium. It is also included on a purely empirical basis in shampoos used for the treatment of dandruff. Hair is often used as an indicator of trace element status of man, and the use of these shampoos makes the selenium analysis of hair worthless, since it does not wash off. Morris et al. (1983) have proposed the use of toenail clippings instead of hair for assaying selenium status. In spite of the widespread industrial uses of selenium, cases of human toxicosis, both of industrial as well as of food origin, are very rare indeed. This is, of course, much more likely to indicate good industrial practice rather than the non-toxicity of selenium to humans.

Dietary Selenium

Selenium in food is mainly in the organic form, as selenomethionine, selenocysteine and selenocystine. The selenium levels of foods vary widely with the geochemical status of the soil. Liver and kidney are rich sources of selenium, and fish in particular, since they concentrate selenium in the liver, probably to counteract mercury toxicity (Burk, 1983; Combs and Combs, 1984). In addition to mercury, cadmium may interfere with the utilization of selenium. Selenium is excreted in urine as the trimethylselenonium ion, $(CH_3)_3Se^+$ and eliminated in breath as dimethylselenide, $(CH_3)_2Se$, which has a characteristic garlicky smell. This is often an early sign of excessive selenium intake. Chronic selenium toxicity, introduced in experimental animals at dietary levels of 2,000-10,000 ng/g, leads to liver injury, bleeding and anemia.

The estimated normal dietary intake ranges from 60-250 μg/day (Burk, 1984). High values (mean 265 ng/ml, highest 600 ng/ml) have been reported for human blood from South Dakota without any obvious ill effect (Howe, 1979). One would expect that dietary intake and blood levels of selenium were considerably higher in Dakotan residents in the 1930's, since people then consumed much more of their own farm produce. One can guess that the daily selenium intake of Dakotans in the 1930's commonly ranged from 200-2000 μg/day and that their blood levels were 200-2000 ng/ml. For comparison one might add that sheep which died of acute selenium poisoning from eating selenium indicator plants, consumed between 200 and 600 mg Se. The health complaints of the Dakotans then were not very specific, mainly bad teeth (see Lo and Sandi, 1980). Values of up to 800 ng/ml have been reported in Venezuelan children. Selenium intake is very low in some parts of the world, notably China, New Zealand and Finland. Keshan disease, which is an endemic cardiomyopathy mainly affecting children and women in China, occurs in regions poor in selenium, and is responsive to selenium supplementation. Blood levels as low as 10 ng/ml have been reported in Keshan disease patients, and daily dietary intake there is probably less than 30 μg (Chen et al., 1980).

Selenium in New Zealand

Thorough and systematic studies in New Zealand have also revealed very low blood selenium concentration, with mean values near 70 ng/ml, whereas common values in the

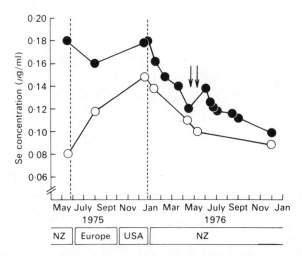

Figure 4. Whole blood selenium concentration (μg/ml) of two New Zealand subjects H (•) and J (o) from May 1975 to December 1976, including period outside New Zealand (NZ), June-December, 1975. Subject H had participated in a study on selenium supplementation in May 1975 and May 1976 (Robinson et al., 1978).

United States and Canada range between 150-250 ng/ml. New Zealand residents travelling abroad show a gradual rise in blood Se while they are away, and the values start falling on their return to New Zealand. This is shown in Figure 4.

John Watkinson who pioneered the fluorimetric determination of selenium in biological material has shown (Watkinson, 1981) that in the North Island of New Zealand, where relatively selenium-rich Australian wheat is occasionally imported to supplement local production, the blood selenium levels of the residents vary with wheat imports. They go up shortly after import of wheat and fall when import ceases. The Southern Island, on the other hand, is self-sufficient in wheat, and no such fluctuations with time have been seen in the residents there, whose blood Se levels are lower than those in the North Island. Watkinson points out that it must be very rare indeed that the dietary level of a single trace element for over half the population of a country depends so much on the import of a single food commodity. No deficiency diseases have been seen in New Zealanders on a normal diet, but a selenium responsive muscle complaint has been described in a New Zealand patient on total parenteral nutrition (van Rij et al., 1979). The selenium status of another New Zealand patient on total parenteral nutrition is shown in Figure 5.

People on restricted diets, like children with phenylketonuria or maple syrup urine disease, have lowered blood selenium, as shown in Figure 6. The selenium status of New Zealanders has been reviewed (Thomson and Robinson, 1980).

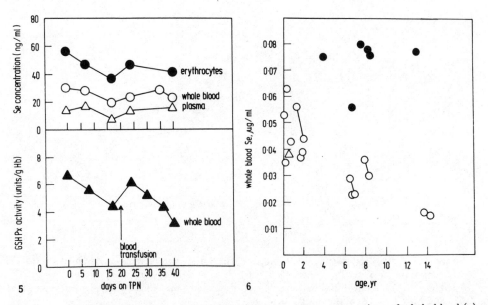

Figure 5. Glutathione peroxidase activity (▲) in blood and Se concentrations of whole blood (o), erythrocytes (•), and plasma (Δ) of a patient during total parenteral nutrition and after a single blood transfusion of 800 ml (van Rij et al., 1979).

Figure 6. Se concentrations in whole blood of infants and children with maple syrup urine disease (MSUD) and phenylketonuria (PKU). PKU synthetic diet (o), PKU nonsynthetic diet (•), MSUD synthetic diet (Δ) [McKenzie et al., 1978].

Selenium and Cancer

Early studies showed that rats fed very high levels of selenium (5,000-10,000 ng Se/g diet) developed cancer. The criticisms made of this study are related by Whanger (1983). At high levels, selenium is certainly toxic and teratogenic. Chicken are extremely sensitive to the teratogenic effects of high selenium levels, a fact of much recent relevance in view of the threat to wildlife at the Kesterson National Wildlife Refuge in California (Marshall, 1985). Other reports have indicated that fairly low levels of selenium can prevent chemically induced cancers in experimental animals (Burk, 1983; Lo and Sandi, 1980). Epidemiological studies suggest an inverse relationship between selenium availability and cancer incidence (Cowgill, 1983; Salonen et al., 1984). Similar and earlier epidemiological studies have been challenged on the basis of data from New Zealand (Thomson and Robinson, 1980), and great care is obviously needed in the interpretation of these data. The rationale behind the hypothesis that selenium can lower the incidence of some types of cancer is that in selenium deficiency, detoxification of harmful metabolites is inadequate. It has been pointed out that selenium may promote carcinogenesis or other toxic effects by increasing the transformation of some xenobiotics (Burk, 1983). When glutathione peroxidase catalyzes the reduction of organic hydroperoxides, alcohols are formed. They are presumed to be harmless, but we wonder if they are metabolically inert? Catalase can use simple alcohols, like methanol and ethanol as substrates. Butanol and propanol are poor substrates, and we would expect the same to apply to the alcohols or hydroxyacids produced by the action of glutathione peroxidase. We wonder if there are other enzymes capable of transforming further the products of the glutathione peroxidase reaction, even to harmful substances?

Selenium and Human Health

Lowered selenium levels and/or glutathione peroxidase activity have been reported in patients with neurological diseases like multiple sclerosis in Denmark, Finland and Israel. Studies in Britain have shown no significant differences between glutathione peroxidase activity of patients and controls, and reports from Poland indicate elevated glutathione peroxidase activity in patients with multiple sclerosis. Our own studies on multiple sclerosis patients and healthy individuals in Iceland (together with J.E.G. Benedikz, G. Gudmundsson and T. Thorsteinsson, Abstracts, 10th Nordic Atomic Spectroscopy and Trace Element Conference, Turku, Finland, 6-9 August 1985, p. 106), revealed no difference in glutathione peroxidase activity, but somewhat higher levels of selenium in whole blood (141 ng/ml) and plasma (109 mg/ml) in multiple sclerosis patients than in normal individuals, who had whole blood levels of 126 ng/ml and plasma Se of 102 ng/ml. The situation is also confused and contradictory with respect to the glutathione peroxidase activity in various types of blood cells of multiple sclerosis patients and healthy people. Glutathione peroxidase activity of erythrocytes or whole blood is actually a poor indicator of human selenium status, contrary to the situation in farm animals. This problem is treated in more detail in Section 3. It has been suggested that the glutathione peroxidase activity of platelets is a reliable indicator of selenium status.

Selenium deficiency has also been suggested to contribute to cardiovascular disease and inflammatory conditions. Selenium therapy has been tried on patients with various diseases in countries where selenium intake is low. The rationale behind such therapy is that it should help to render harmless the peroxides formed by tissue damage caused by the disease. Glutathione peroxidase activity is therefore expected to be more informative than selenium levels in evaluating the need for and the result of such selenium therapy. It would be premature to dismiss selenium therapy as worthless in patients on a low selenium intake, but in those with adequate selenium status it is of doubtful use.

Many of the trace element and vitamin deficiencies seen in livestock result to a greater or lesser extent from intensive farming methods. These same deficiencies are obviously much rarer in humans, since they eat a much more varied diet. For a long time, doubts were raised if vitamin E was needed for humans. Many of the claims for its beneficial effects have indeed not been soundly based. The recommended dietary allowances are 10 mg alpha-tocopherol equivalent per day for men and 8 mg for women. Dietary intake is normally much higher (Bieri, 1984). Neurological syndromes, thought to be caused by vitamin E deficiency, have been described in patients with abetalipoproteinemia and impaired fat absorption. Infants have low vitamin E status, and vitamin E is thought to protect premature babies from retinopathy of prematurity (formerly called

25

retrolental fibroplasia), an eye disease leading to blindness. This disease is associated with the high oxygen tension used in incubators for premature babies (Halliwell and Gutteridge, 1985). It should be pointed out that the benefits of vitamin E in preventing retinopathy of prematurity are doubted by many leading pediatricians (Committee on Fetus and Newborn, 1985). Adverse effects of large doses to such small infants cannot be ruled out. Other factors than high oxygen tension, like bright lights, could also be important, but they are also likely to promote the formation of free radicals (Riley and Slater, 1969; Glass et al., 1985).

The studies of Levander and Morris (1984) indicate that humans need 1 μg dietary selenium per kg body weight per day to maintain balance. A provisional recommended dietary allowance of 50-200 μg Se per day has been given (Burk, 1984). A maximum tolerable level of 500 μg/day has been suggested by Lo and Sandi (1980). Selenium supplementation of vitamin-mineral capsules (40 μg per capsule) has been allowed in some countries. Selenium tablets, containing up to 100 μg each are also freely available. Where selenium intake is high, such supplementation is unnecessary. Many other ingredients of vitamin-mineral tablets are harmless as well as useless to the adequately nourished person. In a few isolated areas the use of such preparations could even be a matter of concern.

The medical profession and human nutritionists have much to learn in this respect from their colleagues in veterinary medicine and animal nutrition. Surprisingly, many of them still subscribe to the "belt and braces hypothesis" or argue that there is absolutely no need whatsoever for vitamin E in human nutrition. Lack of adequate professional knowledge makes the public an easy prey for quacks and food faddists, who are all too eager to catch ignorant professionals with their pants down.

PART 3. GLUTATHIONE PEROXIDASE: METHODS OF ASSAY AND SOME FUNDAMENTAL PROPERTIES

Fundamentals of Assay Method

Glutathione peroxidase (EC 1.11.1.9) was discovered in erythrocytes by Mills (1957). It was clear it had a role in protecting erythrocyte membranes and hemoglobin against oxidative damage. When it was realized that it was a selenoenzyme (Rotruck et al., 1973), immediate interest arose in assaying its activity to indicate the selenium status of farm animals.

Although the assay of glutathione peroxidase activity is fairly easy and quicker than selenium determinations, it is not free of doubts and difficulties, and thorough understanding of the enzyme and its properties is necessary. Originally, it was assayed by a direct end-point method where consumption of GSH is determined by colorimetry or polarography. These methods are fairly laborious, but the introduction of a coupled assay procedure by Paglia and Valentine (1967) made measurements more convenient. In this procedure, glutathione reductase is used as an auxiliary enzyme to regenerate reduced glutathione from oxidized glutathione at the expense of NADPH. The reaction can then be followed kinetically by measuring the rate of falling of absorbance at 340 nm or 366 nm. Since it is convenient (but expensive) to use high concentrations of NADPH, many workers prefer to use 366 nm. The principle of the coupled assay method is indicated in the diagram below, where GSH-Px stands for glutathione peroxidase and GR for glutathione reductase.

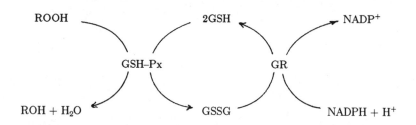

Many tissue extracts have catalase activity ($2H_2O_2 \rightarrow 2H_2O + O_2$), but this is overcome by adding sodium azide, a catalase inhibitor, to the reaction mixture. It would be worth trying aminotriazole as a catalase inhibitor in this assay. The enzyme has a very high specificity for the thiol substrate. On the other hand, it acts on hydrogen peroxide and a great number of organic peroxides, including steroid hydroperoxides and linoleic and linolenic acid hydroperoxides. It does not act on esterified fatty acid hydroperoxides or hydroperoxidized phospholipids, which raises questions about its physiological function as well as the order of events in tissue damage. Cumene hydroperoxide and t-butyl hydroperoxide are the most commonly used organic peroxide substrates used for assaying enzyme activity. High concentrations of glutathione give high blank reaction rates (i.e., the non-enzymic reaction of GSH with peroxides), particularly with hydrogen peroxide. For these reasons, organic hydroperoxides have been preferred as substrates. The apparent k_M values obtained for glutathione are very high and it is impossible to achieve saturation with respect to both substrates at the same time.

Mechanism

The enzyme obeys a ping-pong mechanism which is shown in Figure 7.

The selenolate form of the enzyme first reacts with a hydroperoxide to form a selenenic acid derivative which, in turn, slowly forms a complex with GSH. This newly formed complex is rapidly transformed into the seleno-sulfide derivative of the enzyme, which reacts with a second molecule of GSH, releasing GSSG and the selenolate anion form of the enzyme, which can then start a new catalytic cycle. The observation that it is practically impossible to achieve saturation of the enzyme with respect to glutathione is best explained by the slow formation of a postulated enzyme-glutathione substrate complex from the selenenic acid form of the enzyme, followed by its rapid transformation into a seleno-sulfide derivative. Forms of glutathione peroxidase, where glutathione is covalently attached to the enzyme, have actually been prepared. It is important to have these different forms of the enzyme in mind, when one is investigating the effect of storage conditions and inhibitors (e.g., cyanide).

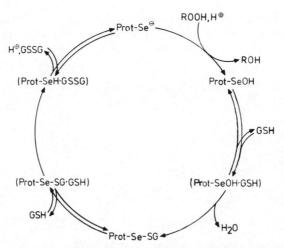

Figure 7. Proposed reaction mechanism of glutathione peroxidase. Prot-Se$^-$, Prot-SeOH, and Prot-Se-SG represent respectively the three different redox states for the enzyme-bound selenium: a selenolate, a selenenic acid and a seleno-sulfide formed between the enzyme and GSH. The expressions in parentheses indicate enzyme-substrate or enzyme product complexes (Günzler et al., 1984).

Can the Assay be Improved?

The final concentrations of GSH used in the several assay methods published vary from 1 mM through 3mM to 6 mM. At high concentrations one naturally has to use more NADPH. Its initial concentration, or more precisely, initial amount, should be sufficient to allow one to follow the reaction long enough to get reliable kinetics. The NADPH concentration should of course not be so high that the spectrophotometer is off scale. The use of 366 nm instead of 340 comes in useful here, although one suspects that much of its continued use in the scientific literature is a hangover from the days of mercury lamps. At higher GSH concentrations, one should also be sure that one is using a suitable excess of the auxiliary enzyme, glutathione reductase. It is not clear if the reported product inhibition of the reductase by GSSG is likely to interfere with the assay. Numerous modifications of the Paglia and Valentine procedure have been published. One has to hope that they are soundly based. To some extent, one can choose the assay conditions in accordance with the performance of the spectrophotometer one is using. Nevertheless, there is a genuine need to optimize the assay conditions with respect to glutathione concentration and the choice of peroxide substrate. The results should naturally be related to the selenium content, if one is using erythrocytes from animals other than humans.

Whole blood or washed erythrocytes have been much used to evaluate the selenium status of animals. The use of whole blood is much more convenient and introduces hardly any error, since the activity of glutathione peroxidase in plasma is very low compared with that of erythrocytes. The effects of anti-coagulants and storage conditions have been studied, but some of the results are conflicting (Agergaard and Jensen, 1982;

Günzler et al., 1974). Much of this confusion may arise from lack of communication between the fundamental enzymologists and the clinical enzymologists. The latter are very often veterinarians who have acquired very good knowledge of enzymology and kinetics, but are hampered by working in relative isolation. Unlike their colleagues in academic hospitals, the staff of veterinary schools and research institutes lack the necessary back-up and stimulation from thriving academic biochemistry and chemistry departments. It would be desirable to undertake a systematic reappraisal of the storage and assay conditions for glutathione peroxidase, taking into account the different oxidation states that exist for the enzyme, which were described in the reaction mechanism.

Another complicating factor is the existence in many tissues of a non-selenium dependent glutathione peroxidase activity, which has been attributed to glutathione S-transferase. This enzyme has not been found in erythrocytes or normal plasma, but it is present in liver, kidney, adrenal, brain and testes (Lawrence and Burk, 1978). This enzyme does not act on hydrogen peroxide, but on many organic peroxides, thus allowing differentiation between the two enzymes. Opinions vary, but glutathione S-transferase is not generally thought to take any major part in peroxide removal *in vivo*, but in this context it is more an analytical problem encountered *in vitro*. Glutathione S-transferase in human plasma has recently been reported to be of use in assessing liver damage (Beckett et al., 1985). It would be interesting to see to what extent this enzyme could contribute to the activity of plasma glutathione peroxidase activity, assayed with organic hydroperoxides.

Selenium, Peroxidases and Heme-Proteins

A further complication arises, since hemoglobin also possesses a peroxidase activity. This problem is particularly acute in the case of human hemoglobin. Hemoglobin is therefore converted to cyanmethemoglobin, but the reagent used for this contains cyanide, which inhibits glutathione peroxidase (Kraus et al., 1980). Transformation of hemoglobin to cyanmethemoglobin may not overcome this problem completely. No suitable inhibitor of this pseudoperoxidase activity seems available. The assay, especially of the human enzyme, would also need a thorough study with respect to this problem. Many of these points have been discussed in review articles (Günzler et al., 1974; Wendel, 1980 and 1981), and by Agergaard and Jensen (1982).

Recent studies by Beilstein and Whanger (1983) have complicated the picture even further, but they may also help to clarify it. They investigated the distribution of selenium and glutathione peroxidase in blood fractions from humans, rhesus and squirrel monkeys, sheep, and rats. They used hydrogen peroxide as substrate and determined selenium by fluorimetry. The majority of selenium in erythrocytes and plasma from rats, sheep, and squirrel monkeys cochromatographed by gel filtration with glutathione peroxidase activity. The elution profiles of sheep plasma and erythrocytes are shown in Figure 8.

Selenium in plasma from humans and rhesus monkeys cochromatographed with two proteins devoid of glutathione peroxidase activity. Nearly all of the selenium in human and rhesus monkey erythrocytes cochromatographed, very surprisingly, with the hemo-

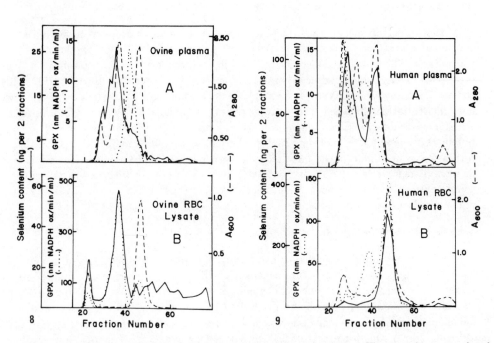

Figure 8. Gel filtration of ovine plasma and lysed erythrocytes. The samples were eluted through a column (2 x 10 cm) of Sephadex G-150 with 0.05 M phosphate buffer, pH 6.3, at a flow rate of 12 ml per hour at 4°. About 6 ml were collected per fraction. The fractions containing hemoglobin were diluted 1 to 5 before absorbance at 600 nm was read. The patterns represent pooled samples from 2 sheep (Beilstein & Whanger, 1983).

Figure 9. Gel filtration of plasma and lysed erythrocytes from humans. Column conditions are same as for Fig. 8. The fractions containing hemoglobin wee diluted 1 to 10 before absorbance at 600 nm was read. The patterns represent pooled samples from 1 man and 2 women (Beilstein & Whanger, 1983).

globin peak. This peak also has significant glutathione peroxidase activity in humans and rhesus monkeys, which also is surprising. The elution profiles of human plasma and erythrocytes are shown in Figure 9.

The hemoglobin peak from squirrel monkey, sheep, and rat blood, on the other hand, contains very little selenium and has hardly any glutathione peroxidase activity. It appears that in man and higher primates, only 10-15% of erythrocyte selenium is bound to glutathione peroxidase, while in most other animals this fraction is 80-90%. This explains why in humans, glutathione peroxidase activity is not a good measure of selenium status, with the exception of people on low selenium intake, like in New Zealand (Rea et al., 1979).

These results should encourage further studies on selenium balance and dynamics in humans. Is this some sort of an overspill mechanism? The results emphasize the importance of thorough comparative work in biochemistry and how risky it is to assume that any enzyme or protein from different species has the same properties. One could speculate that the low molecular weight selenium-binding protein in lamb muscle, absent in selenium-deficient lambs, is related to myoglobin, since this protein absorbs light in the visible region of the spectrum and has been thought to be similar to cytochromes. Ethnic variation in erythrocyte glutathione peroxidase has been described in people of Jewish or Mediterranean origin. Genetic variation in glutathione peroxidase activity of sheep has also been described. Although the only function of selenium known for certain at present is as a part of glutathione peroxidase, other functions can not be excluded, and are in fact quite likely.

Structure, Synthesis and Biomimetic Catalysts

The refined three-dimensional structure of the enzyme (see Figure 10) has been studied by Epp et al. (1983), and the amino acid sequence was elucidated by Günzler et al. (1984).

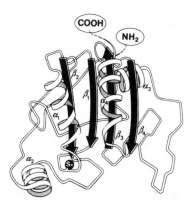

Figure 10. Schematic drawing of the folding pattern of a glutathione peroxidase subunit. Alpha and beta indicate alpha-helices and beta-strands respectively. Se shows the position of the selenocysteine residue.

Epp and his colleagues tried to predict the amino acid sequence from the crystallographic data. This was a somewhat rash thing to do, and it turned out that they had missed a very hydrophobic short peptide at the amino end of the protein. This peptide could function as a hinge whereby the enzyme is attached to hydrophobic parts of membranes. Two forms of glutathione peroxidase from human erythrocytes, both free of hemoglobin, have indeed recently been separated by hydrophobic chromatography (Forward and Almog, 1985). The enzyme is tetrameric and has a molecular weight of 84,000. The identical monomers consist of 198 amino acid residues with the active site selenocysteine located at position 45. Tappel (1984) has provided evidence that there is a selenocysteine-specific aminoacyl t-RNA, but earlier reports favored the interpretation that the selenocysteine residue was formed by post-translational modification.

1986 marks the year in which biochemists will be celebrating the sixtieth anniversary of the first isolation of an enzyme in crystalline form. James B. Sumner, who achieved this literally single-handed (he lost one arm in an accident at the age of seventeen) was at first severely criticized for maintaining that enzymes were proteins by leading figures of the scientific establishment, such as Richard Willstätter, the German Nobel Prize winner. Willstätter claimed that enzymes were catalysts of small molecular weight, associated with an inert colloidal carrier. My colleague, Dr. Hördur Filippusson, has pointed out to me that Willstätter was not entirely wrong. Does not the idea of the active site of an enzyme correspond well with Willstätter's notions? Recently, a selenoorganic compound, PZ 51 (Ebselen), which is shown in Figure 11, has been studied and found to have glutathione peroxidase activity (Müller et al., 1984; Parnham and Kindt, 1984; Wendel et al., 1984).

Epp and his colleagues tried to predict the amino acid sequence from the crystallographic data. This was a somewhat rash thing to do, and it turned out that they had missed a very hydrophobic short peptide at the amino end of the protein. This peptide could function as a hinge whereby the enzyme is attached to hydrophobic parts of membranes. Two forms of glutathione peroxidase from human erythrocytes, both free of hemoglobin, have indeed recently been separated by hydrophobic chromatography (Forward and Almog, 1985). The enzyme is tetrameric and has a molecular weight of 84,000. The identical monomers consist of 198 amino acid residues with the active site selenocysteine located at position 45. Tappel (1984) has provided evidence that there is a selenocysteine-specific aminoacyl t-RNA, but earlier reports favored the interpretation that the selenocysteine residue was formed by posttranslational modification.

When this publication appears in 1986, biochemists will be celebrating the sixtieth anniversary of the first isolation of an enzyme in crystalline form. James B. Sumner, who achieved this literally single-handed (he lost one arm in an accident at the age of seventeen) was at first severely criticized for maintaining that enzymes were proteins by leading figures of the scientific establishment, such as Richard Willstätter, the German Nobel Prize winner. Willstätter claimed that enzymes were catalysts of small molecular weight, associated with an inert colloidal carrier. My colleague, Dr. Hördur Filippusson, has pointed out to me that Willstätter was not entirely wrong. Does not the idea of the active site of an enzyme correspond well with Willstätter's notions? Recently, a selenoorganic compound, PZ 51 (Ebselen), which is shown in Figure 11, has been studied and

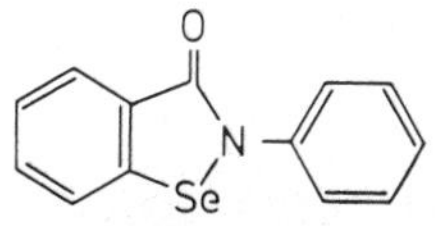

PZ 51

Figure 11. Structure of PZ 51 [Ebselen, 2-phenyl-1,2-benzisoselenazol-3 (2H)-on].

found to have glutathione peroxidase activity (Müller et al., 1984; Parnham and Kindt, 1984; Wendel et al., 1984).

It has no effect on its own on peroxides; it needs glutathione. Its sulfur analogue is nearly completely devoid of any catalytic activity. Copper diisopropyl salicylate (Cu-DIPS) is another biomimetic catalyst with superoxide dismutase activity, but is poorly absorbed from the intestine. PZ 51, on the other hand, is lipophilic, and enters the body easily. Studies of such biomimetic catalysts could offer a new approach to therapy, as well as throw light on the catalytic mechanism and the active site of the enzyme.

REFERENCES

Agergaard, N., and Jensen, P.T., 1982, Procedure for blood glutathione peroxidase determination in cattle and swine, *Acta Vet. Scand.*, 23:515.

Ammerman, C.B., and Miller, S.M., 1975, Selenium in ruminant nutrition: A review, *J. Dairy Sci.*, 58:1561.

Anderson, P.H., and Patterson, D.S.P., 1976, Nutritional myopathy of young cattle, in "Proceedings, Roche Symposium for the Feed Industry," Central Veterinary Laboratory, Weybridge.

Anderson, P.H., Berrett, S., Catchpole, J., and Gregory, M.W., 1983, Effect of monensin on the selenium status of sheep, *Vet. Rec.*, 113:498.

Beckett, G.J., Chapman, B.J., Dyson, E.H., and Hayes, J.D., 1985, Plasma glutathione S-transferase measurements after paracetamol overdose: evidence for early hepatocellular damage, *Gut*, 26:26.

Beilstein, M.A., and Whanger, P.D., 1983, Distribution of selenium and glutathione peroxidase in blood fractions from humans, rhesus and squirrel monkeys, rats and sheep, *J. Nutr.*, 113:2138.

Bergen, W.G., and Bates, D.B., 1984, Ionophres: Their effect on production efficiency and mode of action, *J. Anim. Sci.*, 58:1465.

Bieri, J.G., 1984, Vitamin E, in "Present Knowledge in Nutrition," 5th edition, The Nutrition Foundation, Washington, D.C.

Burk, R.F., 1983, Biological activity of selenium., *Ann. Rev. Nutr.*, 3:53.

Burk, R.F., 1984, Selenium, in "Present Knowledge in Nutrition," 5th edition, The Nutrition Foundation, Washington, D.C.

Chen, X., Yang, G., Chen, J., Chen, X., Wen, Z., and Ge, K., 1980, Studies on the relations of selenium and Keshan disease, *Biol. Trace Elem. Res.*, 2:91.

Combs, G.F., Jr., and Combs, S.B., 1984, The nutritional biochemistry of selenium, *Ann. Rev. Nutr.*, 4:257.

Committee on Fetus and Newborn, 1985, Vitamin E and the prevention of retinopathy of prematurity, *Pediatrics*, 76:315.

Cowgill, U.M., 1983, The distribution of selenium and cancer mortality in the continental United States, *Biol. Trace Elem. Res.*, 5:345.

Eiriksdóttir, G., Símonarson, B., Thorsteinsson, T., Gudmundsson, B., and Jónmundsson, J.V., 98, Árstíðabundnar breytingar á seleni í blóði saudfjár. Tilraun á Hvanneyri 1980, *Ísl. Landbún.*, 13:25 (with English summary).

Epp, O., Ladenstein, R., and Wendel, A., 1983, The refined structure of the selenoenzyme glutathione peroxidase at 0.2-nm resolution, *Eur. J. Biochem.*, 1133:51.

Fishbein, L., 1983, Environmental selenium and its significance, *Fundam. Appl. Toxicol.*, 3:411.

Forward, R., and Almog, R., 1985, Separation of two forms of glutathione peroxidase from human erythrocytes by hydrophobic chromatography, *J. Chromatogr.*, 330:383.

Glass, P., Avery, G.B., Subramanian, K.N.S., Keys, M.P., Sostek, A.M., and Friendly, D.S., 1985, Effect of bright light in the hospital nursery on the incidence of retinopathy of prematurity, *New Engl. J. Med.*, 313:401.

Günzler, W.A., Kremers, H., and Flohé, L., 1974, An improved coupled test procedure for glutathione peroxidase (EC 1.11.1.9.) in blood, *z. Klin. Chem. Klin. Biochem.*, 12:444.

Günzler, W.A., Steffens, G.J., Grossmann, A., Kim, S.-M.A., Otting, F., Wendel, A., and Flohé, L., 1984, The amino acid sequence of bovine glutathione peroxidase, *Hoppe-Seyler's Z. Physiol. Chem.*, 365:195.

Halliwell, B., and Gutteridge, J.M.C., 1985, "Free Radicals in Biology and Medicine," Clarendon Press, Oxford.

Hoekstra, W.G., 1974, Biochemical role of selenium, in "Trace Element Metabolism in Animals - 2," Hoekstra, W.G., Suttie, J.W., Ganther, H.E., and Mertz, W., eds., University Park Press, Baltimore.

Hoekstra, W.G., 1975, Biochemical function of selenium and its relation to vitamin E., *Fed. Proc.*, 34:2083.

Howe, M., 1979, Selenium in the blood of South Dakotans, *Arch. Environ. Health*, 34:444.

Kraus, R.J., Prohaska, J.R., and Ganther, H.E., 1980, Oxidized forms of ovine erythrocyte glutathione peroxidase. Cyanide inhibition of a 4-glutathione:4-selenoenzyme, *Biochim. Biophys. Acta*, 65:19.

Låg, J., and Steinnes, E., 1978, Regional distribution of selenium and arsenic in humus layers of Norwegian forest soils, *Feoderma*, 20:3.

Lawrence, R.A., and Burk, R.F., 1978, Species, tissue and subcellular distribution of non Se-dependent glutathione peroxidase activity, *J. Nutr.*, 108:211.

Levander, O.A., and Morris, V.C., 1984, Dietary selenium levels needed to maintain balance in North American adults consuming self-selected diets, *Am. J. Clin. Nutr.*, 39:809.

Lo, M.-T, and Sandi, E., 1980, Selenium: Occurrence in foods and its toxicological significance - a review., *J. Environ. Pathol. Toxicol.*, 4:193.

Marshall, E., 1985, Selenium poisons refuge, Californian politics, *Science*, 229:144.

McKenzie, R.L., Rea, H.M., Thomson, C.D., and Robinson, M.F., 1978, Selenium concentration and glutathione peroxidase activity in blood of New Zealand infants and children, *Am. J. Clin. Nutr.*, 31:1413.

McMurray, C.H., and Blanchflower, W.J., 1979, Application of a high-performance liquid chromatographic fluorescence method for the rapid determination of α-tocopherol in the plasma of cattle and pigs and its comparison with direct fluorescence and high-performance liquid chromatography-ultraviolet detection methods, *J. Chromatogr.*, 178:525.

McMurray, C.H., Rice, D.A., and Kennedy, S., 1983, Experimental models for nutritional myopathy, in "Biology of Vitamin E," Ciba Foundation Symposium 101, Pitman, London.

Mills, G.C., 1957, Hemoglobin catabolism. I. Glutathione peroxidase, an erythrocyte enzyme which protects hemoglobin from oxidative breakdown, 2J. Biol. Chem., 229:189.

Morris, J.S., Stampfer, M.J., and Willett, W., 1983, Dietary selenium in humans. Toenails as an indicator, Biol. Trace Elem. Res., 5:529.

Müller, A., Cadenas, E., Graf, P., and Sies, H., 1984, A novel biologically active seleno-organic compound - I. Glutathione peroxidase-like activity in vitro and anti-oxidant capacity of PZ 51 (Ebselen), Biochem. Pharmac., 33:3235.

Paglia, D.E., and Valentine, W.N., 1967, Studies on the quantitative and qualitative characterization of erythrocyte glutathione peroxidase, J. Lab. Clin. Med., 70:158.

Parnham, M.J., and Kindt, S., 1984, A novel biologically active seleno-organic compound - III. Effects of PZ 51 (Ebselen) on glutathione peroxidase and secretory activities of mouse macrophages, Biochem. Pharmac. 33:3247.

Piwonka, J., Kaiser, G., and Tölg, G., 1985, Determination of selenium at ng/g- and pg/g-levels by hydride generation atomic absorption spctrometry in biotic materials, Fresenius' Z. Anal. Chem., 321:225.

Rea, H.M., Thomson, C.D., Campbell, D.R., and Robinson, M.F., 1979, Relationship between erythrocyte selenium concentrations and glutathione peroxidase (EC 1.11.1.9) activities of New Zealand residents and visitors to New Zealand, Br. J. Nutr., 42:201.

Reddy, C.C., and Massaro, E.J., 1983, Biochemistry of selenium: A brief overview, Fundam. Appl. Toxicol., 3:431.

Riley, P.A., and Slater, T.F., 1969, Pathogenesis of retrolental fibropasia, Lancet, ii:265.

Robinson, M.F., Rea., H.M., Friend, G.M., Stewart, R.D.H., Snow, P.C., and Thomson, C.D., 1978, On supplementing the selenium intake of New Zealanders. 2. Prolonged metabolic experiments with daily supplements of selenomethionine, selenite and fish, Br. J. Nutr., 39:589.

Rotruck, J.T., Pope, A.L., Ganther, H.G., Swanson, A.B., Hafeman, D.G., and Hoekstra, W.G., 1973, Selenium: Biochemical role as a component of glutathione peroxidase, Science, 179:588.

Salonen, J.T., Alfthan, G., Huttunen, J.K., and Puska, P., 1984, Association between serum selenium and the risk of cancer, Am. J. Epidemiol. 120:342.

Schwarz, K., and Foltz, C.M., 1957, Selenium as an integral part of factor 3 against dietary necrotic liver degeneration, J. Am. Chem. Soc., 79:3292.

Tappel, A.L., 1984, Selenium-glutathione peroxidase: Ptroperties and synthesis, in "Current Topics in Cellular Regulation," Vol. 24, DeLuca, M., Lardy, H.A., and Cross, R.L., eds., Academic Press, New York.

Thomson, C.D., and Robinson, M.F., 1980, Selenium in human health and disease with emphasis on those aspects peculiar to New Zealand, Am. J. Clin. Nutr., 33:303.

Underwood, E.J., 1977, "Trace Elements in Human and Animal Nutrition," 4th edition, Academic Press, New York.

van Rij, A.M., Thomson, C.D., McKenzie, J.M., and Robinson, M.F., 1979, Selenium deficiency in total parenteral nutrition, Am. J. Clin. Nutr., 32:2076.

van't Hoff, J.H., 1967, "Imagination in Science", Springer-Verlag, Berlin.

Watkinson, J.H., 1966, Fluorimetric determination of selenium in biological material with 2,3-diaminonaphthalene, Anal. Chem., 38:92.

Watkinson, J.H., 1981, Changes of blood selenium in New Zealand adults with time and importation of Australian wheat, Am. J. Clin. Nutr., 34:936.

Wendel, A., 1980, Glutathione peroxidase, in "Enzymatic Basis of Detoxication," Jakoby, W.B., ed., Academic Press, New York.

Wendel, A., 1981, Glutathione peroxidase, *Methods Enzymol.*, 77:325.

Wendel, A., Fausel, M., Safayhi, H., Tiegs, G., and Otter, R., 1984, A novel biologically active seleno-organic compound - II. Activity of PZ 51 in relation to glutathione peroxidase, *Biochem. Pharmac.*, 33:3241.

Whanger, P.D., 1983, Selenium interactions with carcinogens, *Fundam. Appl. Toxicol.*, 3:424.

SUGGESTIONS FOR FURTHER READING

Biology of Vitamin E, Ciba Foundation Symposium 101, 1983, Pitman, London.

Chance, B., Sies, H., and Boveris, A., 1979, Hydroperoxide metabolism in mammalian organs, *Physiol. Rev.*, 59:527.

Flohé, L., 1979, Glutathione peroxidase: fact and fiction, in "Oxygen Free Radicals and Tissue Damage," Ciba Foundation Symposium 65, Excerpta Medica, Amsterdam.

Rosenfeld, I., and Beath, O.A., 1964, "Selenium. Geobotany, Biochemistry, Toxicity and Nutrition," Academic Press, New York.

Zingaro, R.A., and Cooper, W.C., eds., 1974, "Selenium," Van Nostrand Reinhold Company, New York.

FORMATION AND PROPERTIES OF OXYGEN RADICALS

Klaus-Dieter Asmus

Hahn-Meitner-Institut
Bereich Strahlenchemie
Glienicker Str. 100
D-1000 Berlin 39, FRG

FORMATION AND DETECTION OF OXYGEN RADICALS WITH RADIATION CHEMICAL METHODS

General Introduction

The principle oxygen centered radicals are: $\cdot OH$, $O^-\cdot$, $O_2^-\cdot$, $HO_2\cdot$, $O_3^-\cdot$, $ROO\cdot$, $RO\cdot$. Most of these species have been found or are discussed in connection with biological and toxicological processes. In order to understand their potential action it is necessary to know about their modes of formation and their fundamental chemical properties (Packer, 1984; Bors et al., 1984).

Since radicals are generally rather short-lived, and this is true also for the above oxygen radicals, their study requires either fast enough techniques for direct detection or has to rely on deductions from reaction products or specific product patterns. A very convenient way to generate and investigate free radicals has proven to be Radiation Chemistry. In fact, most of the information available on the physico-chemical properties of oxygen centered radicals — and discussed in this article — stems from radiation chemical experiments.

Radiation Chemical Background

The energy required to form free radicals from molecular compounds is in Radiation Chemistry provided by either high energy particles (e^-, p^+, α, etc.) or by electromagnetic waves (γ-rays, synchrotron radiation, etc.). The mode of energy dissipation in matter shall not be discussed here in any detail, and the reader is referred to the special literature on this topic (Swallow, 1973). For our further understanding it is important, however, to recognize that the high energy (typically MeV) of an incident particle or electromagnetic wave ultimately results in a large number of small energy deposits (only 60-100 eV on average), each of which provides just enough energy for ionizations and excitations of (again on average) one or two molecules. As a result, electrons, charged and neutral radicals as well as non-radical species, are generated. In irradiated water or aqueous solutions these are:

$$e_{aq}^-, \ H\cdot, \ \cdot OH, \ H_{aq}, \ H_2, \ H_2O_2$$

It is further important to realize that the entire process which leads to the formation of these so-called primary species is over in a time much less than 10^{-10} seconds, i.e., before any normal chemistry sets in.

A problem with respect to a selective study of a particular species is, of course, that all the primary species are generated simultaneously with yields in the same order of magnitude. The non-radical species fortunately do not usually interfere with free radical processes on the time scale of pulse radiolysis owing to their comparatively much lower reactivity. This still leaves all the radical species as potential reactants. In aqueous solutions these are the hydrated electron e_{aq}, the hydrogen atom $H\cdot$, and the hydroxyl radical $\cdot OH$, which are formed with yields of G = 2.75,0.6, and 2.8, respectively (The radiation chemical yields are generally expressed in terms of G = number of species per 100 eV absorbed energy; G = 1 is identical to ca. 0.1 μ moles per J absorbed energy). There are, however, a number of reactions by which these primary radicals can selectively be scavenged or even be interconverted. The most important among these processes with respect to the oxygen radicals are the conversion of hydrated electrons into hydroxyl radicals

$$e_{aq}^- + N_2O + H_2O \rightarrow \cdot OH + OH^- + N_2 \tag{1}$$

and the generation of superoxide via

$$e_{aq}^- + O_2 \rightarrow O_2^-\cdot \tag{2}$$

and

$$H\cdot + O_2 \rightarrow HO_2\cdot \tag{3}$$

Other significant reactions concerning the selectivity in an irradiated aqueous solution are the conversion of hydrated electrons into hydrogen atoms

$$e_{aq}^- + H_{aq}^+ \rightarrow H\cdot \tag{4}$$

or the removal of hydrogen atoms and hydroxyl radicals by alcohols

$$H\cdot/\cdot OH + ROH \rightarrow H_2/H_2O + ROH(-H)\cdot \tag{5}$$

In summary: Radiation Chemistry provides the possibility of generating free radicals which in high polarity solvents are mainly homogeneously distributed and whose physical and chemical properties can selectively be studied in appropriately composed systems (Packer, 1984).

Detection of Radicals

The detection of oxygen radicals generated by chemical processes is possible by direct as well as by indirect methods. A direct method is, for example, pulse radiolysis which belongs to the group of time-resolved techniques. Its principles and details are extensively explained in the literature and many textbooks (Swallow, 1973; Baxendale and Busi, 1982; Asmus, 1984). Only the essential shall be mentioned here. The main advantage of pulse radiolysis is that it allows generation of fairly high radical concentrations within a time period (pulse) which is usually short compared to the lifetime of the radicals (typical radical concentrations per pulse are in the 1-10 μmolar range). Detection of the radicals produced as a result of an incident pulse of high energy electrons is achieved by monitoring the time dependence of a physical property such as optical absorption (colour), conductance (charge), spin density, etc. of the radical species. The most commonly used technique is optical detection since many radicals display absorption spectra in the near UV and VIS range. In modern pulse radiolysis, submicromolar radical concentrations are sufficient for detection, i.e., concentrations which can easily be generated in a pulse as short as a few nanoseconds.

A typical example for the application of pulse radiolysis is the reaction

$$e_{aq}^- + O_2 \rightarrow O_2^-\cdot \tag{2}$$

The hydrated electron itself is a highly coloured species ($\gamma_{max} = 720$ nm, $\epsilon = 1.8 \times 10^4$ M^{-1} cm^{-1}). Its concentration-time profile shall be given by the solid line in Fig. 1a. If oxygen is present in the solution, one finds an increasingly faster decay of the signal with increasing O_2 concentration. Since $O_2^-\cdot$ also exhibits an absorption with $\gamma_{max} = 245$ nm [$\epsilon \approx 2000$ M^{-1} cm^{-1} (Behar et al., 1970)], its corresponding formation may be traced in the UV. Analysis of the kinetic curves provides the associated absolute rate constants, e.g., $k = 2.0 \times 10^{10}$ M^{-1} s^{-1} for the superoxide anion formation via reaction 2.

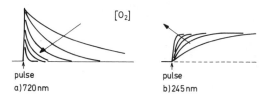

Not all of the radicals exhibit a directly and easily detectable physical property as is the case, for example, of the $\cdot$OH radical. In order to evaluate its reactivity, one has therefore to rely on the observation of a reaction product, e.g., the hydroxycyclohexadienyl radical from the reaction

$$\tag{6}$$

which absorbs in the UV with $\gamma_{max} = 310$ nm. If, however, the reaction product itself also does not exhibit a detectable absorption as is the case, for example, of the alkyl radical formed via

$$\cdot OH + RH \rightarrow R\cdot + H_2O, \tag{7}$$

a competitive method has to be applied. This indirect method is based on the evaluation of the effect of a compound whose reaction cannot be monitored directly (e.g., reaction 7) on a detectable process (e.g., reaction 6). In our example, the observable yield of the hydroxycyclohexadienyl radical, which absorbs at 310 nm, will decrease upon increasing alkane concentration, or better even increasing the $[RH]/[C_6H_6]$ ratio as schematically shown in Fig. 2.

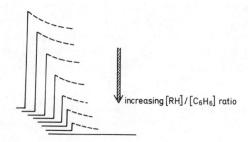

increasing $[RH]/[C_6H_6]$ ratio

The mathematical evaluation of a competitive scheme leads to the general equation

$$Y = Y_o \frac{k_1 [A]}{k_1[A] + k_2[B]}$$

where Y and Y_0 are the measurable yields (here the 310 nm absorption) of the product from compound A (here C_6H_6) in the presence and the absence of the competing compound B (here RH), respectively. A rearrangement of this equation giving

$$\frac{Y_o}{Y} = 1 + \frac{k_2 [B]}{k_1 [A]}$$

shows that the rate constant ratio k_2/k_1 can be obtained from the slope of Y_0/Y vs. $[B]/[A]$ plot. With a known k_1 (which can generally be measured directly), the unknown and not directly measurable rate constant k_2 can then be evaluated.

Despite the availability of the pulse radiolysis technique (and comparable methods like flash photolysis etc.) many investigations have still to rely on conclusions and deductions from the analysis of stable reaction products. Although this conventional method can often give only a much more limited insight into the individual steps of a reaction mechanism, it is nevertheless an important supporting and control factor for the conclusions and mechanisms postulated from the direct observation of the radical intermediates.

40

FORMATION AND CHEMICAL PROPERTIES OF OXYGEN RADICALS

·OH Radicals

Hydroxyl radicals are among several primary species formed in the radiolysis of water, i.e.

$$H_2O \rightsquigarrow \cdot OH + e_{aq}^- + \cdots \tag{8}$$

They are also generated in the reaction of hydrated electrons with nitrous oxide (see reaction 1) and via dissociative electron capture by hydrogenperoxide, e.g.

$$e_{aq}^- + H_2O_2 \rightarrow \cdot OH + OH^- \tag{9}$$

The electrons necessary for the latter process may also be provided by conventional chemical reductants such as metal ions

$$M^{n+} + H_2O_2 \rightarrow \cdot OH + OH^- + M^{(n+1)+} \tag{10}$$

with $M^{n+} = Fe^{2+}$, Ce^{3+}, Ti^{3+} etc. Reaction 10 is known as a Fenton-type reaction.

Hydroxyl radicals undergo various types of reactions, namely one-electron oxidation, addition, and abstraction or substitution processes.

The oxidizing property is reflected in a high oxidation potential which amounts to +1.9 V for the $\cdot OH/OH^-$ couple (+2.7 V for $\cdot OH, H^+/H_2O$). Examples for such one-electron processes are

$$\cdot OH + Fe(CN)_6^{4-} \rightarrow Fe(CN)_6^{3-} + OH^- \tag{11}$$

and

$$\cdot OH + RSSR \rightarrow (RSSR)^{\cdot +} + OH^- \tag{12}$$

The latter process is, however, not the exclusive route for the reaction of $\cdot OH$ with a disulfide. Depending on the nature of R and the pH of the solution

$$\cdot OH + RSSR \left\{ \begin{array}{l} RSO\cdot + RSH \quad\quad (13) \\ RS\cdot + RSOH \quad\quad (14) \end{array} \right.$$

are competing processes. Reaction 14 constitutes a radical displacement while the products of reaction 13 require some more complex molecular rearrangement (Asmus, 1983).

Although its high oxidation potential renders $\cdot OH$ an excellent candidate for direct one-electron oxidations, only a limited number of such reactions have been observed. The main reason for this is another important property of the $\cdot OH$ radical, namely its electrophilicity. This means that $\cdot OH$ is easily added to any electron rich center, e.g., to aromatic systems as in reaction 6. In general, any carbon-carbon double bond will suffer this addition process

$$\begin{array}{c}\diagup\\[-2pt]C=C\\[-2pt]\diagup\quad\diagdown\end{array}\quad+\quad\cdot OH\quad\rightarrow\quad\begin{array}{c}OH\\|\quad\;\cdot\\-\,C-C-\\|\quad\;|\end{array}$$

thereby forming a β-hydroxy radical. Even a free electron pair at a hetero-atom provides a suitable site for an ·OH addition. An example for this is found in the ·OH induced oxidation of a sulfide function, i.e.

$$R-\underset{\cdot\cdot}{\overset{\cdot\cdot}{S}}-R \;+\; \cdot OH \;\rightarrow\; R_2S(OH)\cdot \tag{16}$$

Addition complexes are also frequently found with metal ions, e.g.

$$Tl^+ + \cdot OH \rightarrow Tl(OH)^+ \tag{17}$$

All these hydroxylated radicals and ionic species may subsequently eliminate OH⁻ ions in a process which is generally acid catalysed, e.g.

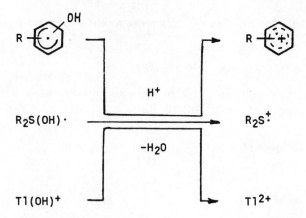

The result of the combined addition-elimination processes is eventually the same as that of straight forward one-electron oxidation. It has to be borne in mind, however, that the ·OH-adducts exhibit chemical characteristics of their own. Prior to their ionization via reaction 18, they may thus channel a reaction into a competing pathway. Generally, the ·OH-adducts have a lower oxidation (or higher reduction) potential than the one-electron oxidation products.

Another interesting aspect which should be mentioned is that the lifetimes of the ·OH-adducts are not always only a function of the free H^+ concentration, i.e., the pH of the solution. Any proton donor such as $H_2PO_4^-$ or protonated aminogroups ($-NH_3^+$) etc. may provide the necessary proton for the OH⁻ elimination. Such a process can be inter- as well as intra-molecular (Mönig et al., 1985).

The final important mode of ·OH action is hydrogen atom abstraction, e.g.

$$- \overset{\displaystyle |}{\underset{\displaystyle |}{C}} - H \ + \ \cdot OH \ \rightarrow \ - \overset{\displaystyle |}{\underset{\displaystyle |}{C}} \cdot \ + \ H_2O \qquad\qquad (19)$$

or

$$- \overset{\displaystyle \cdot\cdot}{\underset{\displaystyle \cdot\cdot}{S}} - H \ + \ \cdot OH \ \rightarrow \ - \overset{\displaystyle \cdot\cdot}{\underset{\displaystyle \cdot\cdot}{S}} \cdot \ + \ H_2O \qquad\qquad (20)$$

Abstraction from a C-H bond is usually associated with some activation energy and the rate constants are consequently found to be lower than for a merely diffusion controlled reaction such as the H-atom abstraction from a thiol group. C-H bond cleavage may nevertheless be of significance, particularly in large molecules with many of these bonds. This is illustrated in Table 1 which shows the yields of radicals observable in the ·OH in-duced oxidation of various alcohols together with the overall abstraction rate constants. Several principles emerge from these data. First of all, they show that hydrogen atom abstraction occurs at all possible sites. Preference is given, however, to C-H bonds which are activated by a functional group (here -OH) located at the same carbon atom. It can further be noted that abstraction is favoured in the series $-CH > CH_2 > -CH_3$ corresponding to the known trend in C-H bond strengths within these groups. It should finally be mentioned that the yields of O-H cleavage do not necessarily reflect the total ·OH attack at the hydroxyl group. Fast bimolecular and intra-molecular processes of the oxygen centered radicals — which will be discussed in the section on RO· radicals — may leave only a fraction of them for direct detection.

In summary, the ·OH radical is a highly reactive species which is capable of reacting via various mechanisms and with many different reaction sites. It is therefore very unselective and its reaction products may consequently be very diverse.

O⁻· Radical Anions

O⁻· radical anions are found to be formed only in very basic aqueous solutions. This is due to the fact that O·⁻ is the conjugate base to the ·OH radical which itself can consequently be viewed as an acid. The pK of the equilibrium

$$\cdot OH \rightleftarrows O^{-} \cdot + H^{+} \qquad\qquad (21)$$

is 11.9.

Chemically the O·⁻ may, in principle, undergo the same type of reactions as the ·OH radical. However, its negative charge strongly reduces its electrophilicity and conse-quently one-electron oxidations as well as addition reaction will be much less efficient. It mainly undergoes abstraction reactions and is therefore slightly more selective than ·OH.

$O_2^-·$ and HO_2

One of the most prominent radical species in biological sciences is known to be the superoxide anion $O_2^-·$. The most straight forward way of its production is electron

Table 1

alcohol	$k(\cdot OH + alc.)$ $M^{-1} s^{-1}$ (average values from Lit.)	radical	%
CH_3OH	8×10^8	$\dot{C}H_2OH$	93
		$CH_3O \cdot$	7
CH_3CH_2OH	1.8×10^9	$CH_3\dot{C}HOH$	84
		$CH_3CH_2O \cdot$	2-3
		$\dot{C}H_2CH_2OH$	13
$CH_3CH_2CH_2OH$	2.7×10^9	$CH_3CH_2\dot{C}HOH$	53
		$CH_3CH_2CH_2O \cdot$	< 1
		$CH_3\dot{C}HCH_2OH$	} 46
		$\dot{C}H_2CH_2CH_2OH$	
$(CH_3)_2CHOH$	2.2×10^9	$(CH_3)_2\dot{C}OH$	86
		$(CH_3)_2CHO \cdot$	1
		$\dot{C}H_2CH(CH_3)OH$	13
$(CH_3)_3COH$	5×10^8	$(CH_3)_3CO \cdot$	4
		$\dot{C}H_2C(CH_3)_2OH$	96

(Identification is possible through the redox properties; α-hydroxy radicals are reducing, oxy radicals oxidizing, the others redox inert).
Lit.: NBRDS series on rate constants (Nat. Bureau of Standards);
K.-D. Asmus, H. J. Möckel, A. Henglein, J. Phys. Chem. 77, 1218 (1973) [A-5]

attachment to molecular oxygen, e.g., in irradiated aqueous solution:

$$e_{aq}^- + O_2 \rightarrow O_2^- \cdot \tag{2}$$

As may be anticipated, this process is very fast and occurs with a bimolecular rate constant of 2.0×10^{10} $M^{-1}s^{-1}$.

Like many other hetero-atom centered radicals (see $\cdot OH/O^- \cdot$), $O_2^- \cdot$ is also involved in an acid-base equilibrium, namely

$$HO_2 \cdot \rightleftarrows O_2^- \cdot + H^+ \tag{22}$$

the pK of which is 4.9. The $HO_2 \cdot$ radical may thus be generated by acidification of an $O_2^- \cdot$ containing system, but also directly via reaction of a hydrogen atom with molecular oxygen

$$H \cdot + O_2 \rightarrow HO_2 \cdot \tag{3}$$

which occurs with $k = 2 \times 10^{10}$ $M^{-1}s^{-1}$.

Both $O_2^- \cdot$ and $HO_2 \cdot$ are relatively easily detectable in pulse radiolysis studies owing to their optical absorptions in the UV. The spectral characteristics as reproduced from a paper by Behar et al. (1970) are shown in the following figure.

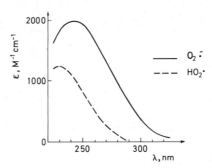

Absorption maxima are found at 245 nm for $O_2^- \cdot$ ($\epsilon \approx 2000$ $M^{-1}cm^{-1}$), and at 230 nm for $HO_2 \cdot$ ($\epsilon \approx 1250$ $M^{-1}cm^{-1}$).

Two further reactions by which superoxide can be generated are an electron transfer from the carbon dioxide radical anion to molecular oxygen

$$CO_2^- \cdot + O_2 \rightarrow O_2^- \cdot + CO_2 \tag{23}$$

($k = 2 \times 10^9$ $M^{-1}s^{-1}$) and via the decay of an α-hydroxy-peroxyl radical

$$\begin{matrix} \text{OH} \\ | \\ -\ \text{C}\ -\ \text{OO}\cdot \\ | \end{matrix} \rightarrow \text{HO}_2\cdot\ +\ \diagdown\!\!\text{C}=\text{O}\diagup \tag{24}$$

($CO_2^-\cdot$ is formed by reduction of CO_2 but also by $\cdot OH$ reaction with formate, $HCOO^-$. The α-hydroxy-peroxyl radical results from O_2 addition to the carbon centered α-hydroxy radical $\ddot{C}OH$).

Chemically, the $O_2^-\cdot$ acts mainly as a mild reductant, for example

$$O_2^-\cdot + C(NO_2)_4 \rightarrow C(NO_2)_3^- + NO_2 + O_2 \tag{25}$$

($k = 2 \times 10^9\ M^{-1}s^{-1}$). An interesting system is established in solutions containing both molecular oxygen and quinones, Q, upon reduction of one of these solutes, namely the equilibrium

$$O_2^-\cdot + Q \rightleftarrows O_2 + Q^-\cdot \tag{26}$$

with $Q^-\cdot$ being the semiquinone anion. Through this equilibrium it has, incidentally, become possible to evaluate the reduction potential of the $O_2^-\cdot/O_2$ couple to $E_7^\circ = -0.33$ V (ref. to 1 atm O_2; -0.19 V ref. to 1 M O_2) in aqueous solution (Forni and Willson, 1984). It is important to note that the reduction potential seems, however, to depend on the nature of the solvent. In aprotic solution, e.g., CH_3CN, it is lowered to -0.57 V, i.e., $O_2^-\cdot$ becomes a better reductant than in aqueous solution. The solvent dependent action of $O_2^-\cdot$ is particularly well demonstrated in the case of CCl_4, a well known liver toxin. While in aqueous solution, $O_2^-\cdot$ does not react at all with CCl_4, it readily does so in CH_3CN via a process which has been suggested to be a nucleophilic substitution (Roberts et al., 1983).

The $HO_2\cdot$ radical is not as good a reductant as $O_2^-\cdot$; in fact, it rather acts as a mild oxidant. For example, $HO_2\cdot$ barely reacts with tetranitromethane (see eq. 25), but oxidizes, ferrous ions

$$HO_2\cdot + Fe^{2+} + H^+ \rightarrow Fe^{3+} + H_2O_2 \tag{27}$$

This difference in redox properties between $HO_2\cdot$ and $O_2^-\cdot$ is, incidentally, typical for radicals existing in acid-base equilibria, i.e., the acid form is generally a better oxidant/worse reductant than the conjugate base. This finding provides also a reasonable rationale for the relatively high bimolecular rate constant for the reaction

$$HO_2\cdot + O_2^-\cdot \rightarrow O_2 + HO_2^-\ (H_2O_2) \tag{28}$$

which occurs with $k = 1 \times 10^8\ M^{-1}s^{-1}$ compared to $k = 9 \times 10^5\ M^{-1}s^{-1}$ for

$$2HO_2\cdot \rightarrow O_2 + H_2O_2 \tag{29}$$

and $k < 0.35\ M^{-1}s^{-1}$ for

$$2O_2^-\cdot \rightarrow O_2 + O_2^{2-}\ (H_2O_2) \tag{30}$$

Only in reaction 28 is it possible to clearly identify the oxidant ($HO_2\cdot$) and the reductant

($O_2^-\cdot$) while in the dismutation reactions 29 and 30 the $HO_2\cdot$ or $O_2^-\cdot$ have to serve both functions at a time (Czapski, 1984; Bielski, 1984).

Without going into detail, it should finally be mentioned that internal electron transfer (usually from metal ions) in oxygen-containing complexes (e.g., chelates, porphyrins, etc.) could lead to an $O_2^-\cdot$ ion closely associated with the complex which itself adopts an oxidized form, e.g.,

$$Fe^{II}\text{-complex} \underline{\quad\quad} O_2 \leftrightarrow Fe^{III}\text{-complex} \underline{\quad\quad} O_2 \tag{31}$$

$O_3^-\cdot$ and $HO_3\cdot$

The ozonide radical anion $O_3^-\cdot$ can most conveniently be generated by the addition reaction

$$O^-\cdot + O_2 \rightarrow O_3^-\cdot \tag{32}$$

which is certainly supported by the electrophilic character of molecular oxygen. Electron attachment to ozone

$$e_{aq}^- + O_3 \rightarrow O_3^-\cdot \tag{33}$$

or electron transfer from the superoxide anion

$$O_2^-\cdot + O_3 \rightarrow O_2 + O_3^-\cdot \tag{34}$$

also yield this species which can directly be identified through its strong optical absorption ($\lambda_{max} = 430$ nm, $\epsilon = 2 \times 10^3$ $M^{-1}s^{-1}$).

As other radical anions the ozonide can also be viewed as a conjugate base in an acid-base equilibrium

$$HO_3\cdot \rightleftarrows O_3^-\cdot + H^+ \tag{35}$$

for which a pK ≈ 8 has been reported. The $HO_3\cdot$, if formed as such, seems however to be very short-lived, and its unambiguous identification is still controversial. In any case, neutralization of $O_3^-\cdot$ (as well as the reaction of $H\cdot$ atoms with O_3) leads ultimately to the destruction of ozone

$$O_3^-\cdot + H^+ \rightleftarrows \left[HO_3\cdot\right] \rightarrow \cdot OH + O_2 \tag{36}$$

thereby providing another interesting source for hydroxyl radicals (Sehested et al., 1983).

Peroxyl Radicals

The usual pathway to generate peroxyl radicals is via addition of molecular oxygen to a radical site, e.g., to carbon centered radicals

$$-\overset{|}{\underset{|}{C}}\cdot + O_2 \rightarrow -\overset{|}{\underset{|}{C}}-O-O\cdot \tag{37}$$

or sulfur centered radicals

$$- \overset{\cdot\cdot}{\underset{\cdot\cdot}{S}} \cdot \ + \ O_2 \ \rightarrow \ - \overset{\cdot\cdot}{\underset{\cdot\cdot}{S}} - O - O \cdot \tag{38}$$

Several metal ions (Cd^+, Co^+, Cu^+, etc.) have also been found to add oxygen

$$M^{n+} + O_2 \rightarrow MOO^{n+} \tag{39}$$

to yield a peroxyl type complex.

A different source of peroxyl radicals is the reduction of hydroperoxides, e.g., by transition metal ions (Aust et al., 1985), i.e.,

$$ROOH + M^{(n+1)+} \rightarrow ROO \cdot + H^+ + M^{n+} \tag{40}$$

The formation of the carbon peroxyl radicals is generally fast and often controlled only by the diffusion of the reactants, i.e., it occurs with bimolecular rate constants in the $10^9 - 10^{10}$ $M^{-1}s^{-1}$ range. Since O_2 is electrophilic the rate constants depend, however, on the actual electron density at the radical site. For example, the hydroxycyclohexadienyl radical formed in reaction 6 adds O_2 with $k = 5 \times 10^8$ $M^{-1}s^{-1}$. The nitrohydroxycyclohexadienyl radical, on the other hand, undergoes the same reaction with only $k = 2.5 \times 10^6$ $M^{-1}s^{-1}$ owing to the electron-withdrawing effect of the nitro group.

The rate constants for the addition of oxygen to thiyl radicals, $RS \cdot$, seem to be generally 2-3 orders of magnitude below the diffusion limit, i.e., in the $10^7 - 10^8$ $M^{-1}s^{-1}$ range. However, some published values lie higher, and an unambiguous interpretation of the available experimental information is still under discussion. In this context it is certainly interesting to note that O_2 addition to $-O \cdot$ radicals, i.e., to the group-VI-neighbor-atom centered radical is not observed at all (Asmus, 1983).

Peroxyl radicals undergo a large variety of reactions. Detailed and conclusive studies have, however, only been conducted for those from carbon centered radicals; the following discussion shall therefore be confined only to these species.

Like most free radicals, peroxyl radicals may also abstract hydrogen atoms, e.g., from C-H bonds

$$- \overset{|}{\underset{|}{C}} - O - O \cdot \ + \ - \overset{|}{\underset{|}{C}} - H \ \rightarrow \ - \overset{|}{\underset{|}{C}} - OOH \ + \ - \overset{|}{\underset{|}{C}} \cdot \tag{41}$$

to yield a hydroperoxide and a carbon-centered radical. In an oxygen-containing solution, the latter will add O_2 again and thus serve as part of a chain process. The absolute rate constants for reaction 41 are, however, generally very low, typically $< 10^3$ $M^{-1}s^{-1}$. In fact, most often the actual steady state radical concentration is sufficient for a bimolecular radical-radical process in competition with the hydrogen abstraction. It is generally accepted that two peroxyl radicals primarily combine to a tetroxide

$$2 \quad - \overset{|}{\underset{|}{C}}OO\cdot \;\; \rightarrow \;\; [\; -\overset{|}{C}OOOO\overset{|}{\underset{|}{C}} - \;] \tag{42}$$

which is very unstable and breaks down via various pathways (the substituents at carbon can, in principle, be of any kind, i.e., an alkyl group, a hydrogen atom, a halogen atom, etc.). Two types of reactions can be distinguished, namely, concerted and non-concerted mechanisms. Since the former require the presence of at least one hydrogen atom α at the peroxyl group, and occur with maximum efficiency with secondary peroxyl radicals, all the possible reactions shall be exemplified with $R_2CHOO\cdot$ radicals (Zegota et al., 1984).

The concerted mechanism results in cleavage of either O_2 or H_2O_2 from the tentative tetroxide, yielding a ketone plus alcohol or two ketone molecules, respectively:

$$[R_2CHOOOOCHR_2]$$

$$\swarrow a) \qquad\qquad \searrow b) \tag{43}$$

$$O_2 + R_2CO + R_2CHOH \qquad\qquad H_2O_2 + 2\ R_2CO$$

Route 43a is known as the Russell-mechanism and involves a cyclic rearrangement

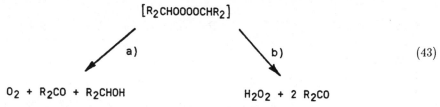

$$\rightarrow \quad R_2CO \;+\; O_2 \;+\; R_2CHOH$$

The non-concerted pathways are based on an oxygen cleavage associated with the generation of free oxy radicals:

$$\left[R_2CHOOOOCHR_2 \right] \rightarrow O_2 + 2R_2CHO\cdot \tag{44}$$

These may combine to form a peroxide

$$2\ R_2CHO\cdot \rightarrow R_2CHOOCHR_2 \tag{45}$$

or disproportionate

$$2\ R_2CHO\cdot \rightarrow R_2CO + R_2CHOH \tag{46}$$

A further possibility is conversion of the oxy radicals into carbon centered α-hydroxy radicals (both via inter- and intra-molecular processes; see next section)

$$R_2CHO\cdot \rightarrow R_2\dot{C}OH \tag{47}$$

In oxygenated solutions, the latter are subsequently subject to O_2 addition to yield an α-hydroxy peroxyl radical. As described already in the case of equation 24, these are then a source of superoxide.

A reaction which applies in particular to tertiary oxy radicals ($=$ no hydrogen atoms left at the α-carbon atom) is radical cleavage

$$R_3CO\cdot \rightarrow R\cdot + R_2CO \tag{48}$$

with R· again being a potential carrier of a chain process.

Particular attention has been given recently to halogenated peroxyl radicals, since they play a significant role in the metabolism of halocarbons. The important aspect is that halogen substituted peroxyl radicals have oxidizing properties. The trichloromethyl peroxyl radical, for example, readily oxidizes electron donors like tyrosine, ascorbate (Vit C), α-tocopherol (Vit E), methionine, etc., in the general reaction

$$CCl_3OO\cdot + D \rightarrow D\cdot^+ + CCl_3OO^- \tag{49}$$

Within a congeneric series of radicals, the oxidation potential increases with increasing halogenation, e.g.

$$CCl_3OO\cdot > CCl_2HOO\cdot > CClH_2OO\cdot > CH_3OO\cdot$$

Typical rate constants for reaction 49 and the above mentioned donors are in the $10^8 M^{-1}s^{-1}$ range, while practically no oxidation is observed through $CH_3OO\cdot$ radicals. The rationale for the oxidizing property is the electronic effect of the chlorine atoms, and it can be anticipated that any other electron-withdrawing substituent at the peroxyl carbon will serve the same function (Packer et al., 1978; Mönig et al., 1983).

Oxy Radicals

Probably the most important source of oxy radicals has already been mentioned in the previous section, namely, the decay of peroxyl radicals. The reaction of hydroperoxides with metal ions

$$-\overset{|}{\underset{|}{C}}-OOH + M^{n+} \rightarrow -\overset{|}{\underset{|}{C}}-O\cdot + OH^- + M^{(n+1)+} \tag{50}$$

(where M^{n+} can be Fe^{2+}, Cu^+, etc.) is another important pathway to generate oxy radicals. A further possibility is cleavage of an O-H bond, e.g., by ·OH radicals

$$-\overset{|}{\underset{|}{C}}-O-H + \cdot OH \rightarrow -\overset{|}{\underset{|}{C}}-O\cdot + H_2O \tag{51}$$

For aliphatic alcohols this is only of minor importance (see section II-1), but for phenols and hydroquinones, i.e., for aromatic systems in which the oxygen centered radical can be resonance stabilized, reaction 51 may account quantitatively for the action of the hydroxyl radicals (The actual mechanism with the aromatic compounds mostly proceeds via an ·OH addition/H_2O elimination sequence for reasons discussed already in the ·OH section above).

The chemistry of the oxy radicals depends strongly on the functional groups attached to the oxygen atom. Aliphatic oxy radicals generally exhibit oxidizing properties, e.g., in

$$RO\cdot + I^- \rightarrow I\cdot + RO^- \tag{52}$$

but also easily undergo hydrogen atom abstraction . If formulated with an alcohol

$$RO\cdot \;+\; -\overset{\overset{\displaystyle H}{|}}{\underset{|}{C}} - OH \;\rightarrow\; ROH \;+\; -\overset{\displaystyle \cdot}{\underset{|}{C}} - OH \tag{53}$$

this means a conversion of an oxidizing oxygen centered radical into a reducing carbon centered radical. Reaction 53, incidentally, is generally quite fast and explains why in alcohols usually only a small yield of oxy radicals is detectable (see section II-1). The latter finding is additionally based on an intramolecular 1,2-hydrogen shift in α-hydrogen substituted oxy radicals

$$-\overset{\overset{\displaystyle H}{|}}{\underset{|}{C}} - O\cdot \;\xrightarrow{\;\;\Omega\;\;}\; -\overset{\displaystyle \cdot}{\underset{|}{C}} - OH \tag{54}$$

which typically seems to occur within the μs time range.

Phenolic oxy radicals are also found to act as oxidants or via hydrogen atom abstraction. Such a process (probably via one-electron oxidation) applies, for example, for the oxidation of Vit C (AH$^-$) by the oxy radical from α-tocopherol

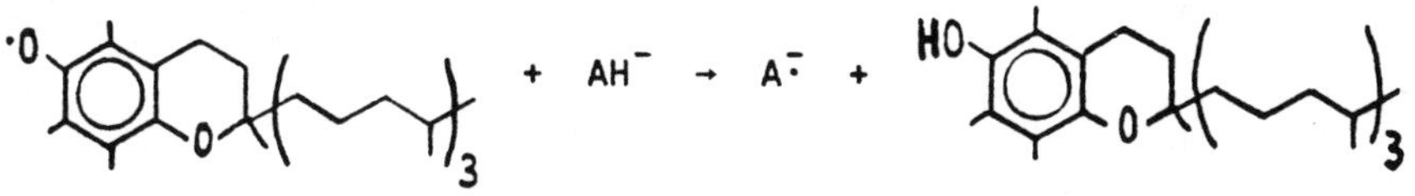

(55)

[k = 1.55 x 10^6 M^{-1}s^{-1} (Packer et al., 1978)].

Free semiquinones are oxy radicals which exhibit acid-base equilibria, e.g.

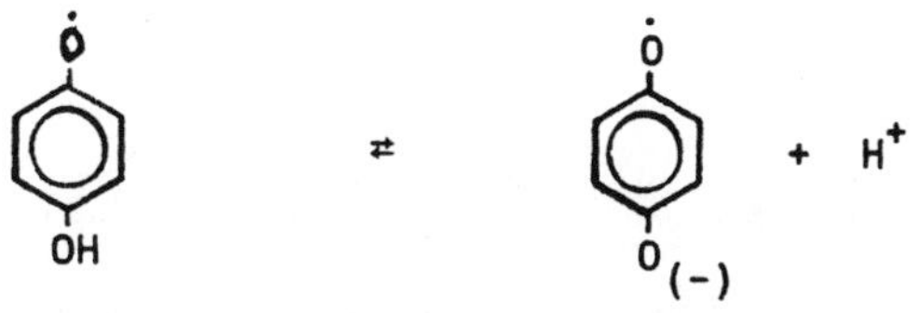

(56)

(QH·)   (Q$^{\bar{\cdot}}$)

[pK = 4.0 for equilibrium 56]. They generally act as reductants because they can easily be oxidized to the corresponding quinones. As example is the dissociative electron transfer to tetranitromethane

$$Q^{-\cdot} + C(NO_2)_4 \rightarrow C(NO_2)_3^- + NO_2 + Q \qquad (57)$$

[$k = 7.11 \times 10^7\,M^{-1}s^{-1}$ for 11,4-semiquinone and $1.3 \times 10^8\,M^{-1}s^{-1}$ for 2-methyl-1,4-semiquinone]. Biologically most important is, of course, the establishment of a redox equilibrium with molecular oxygen as described already by eq. 26 above.

"Krypto"-Oxy Radicals

A free radical by definition is a species with an unpaired electron or spin. These electrons are not necessarily localized. In fact, they seek any possibility for delocalization, since the establishment of such resonance structures is associated with an energy gain for the system. Spin density may thus be spread over aromatic rings, substituents and hetero-atoms. A nitroxyl radical, for example, shows radical reactions both at nitrogen and oxygen.

are only resonance forms of a system which may probably best be viewed as

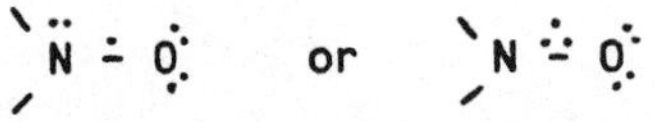

i.e., as species with partial Π-bond between N and O. Another instructive radical is the $(CH_2CHO)\cdot$ radical. It has pronounced oxidizing properties which are typical for an oxy radical. ESR measurements, on the other hand, reveal that the spin of this radical is mainly localized at the methylene-carbon atom. Both observations suggest resonance between the two following structures

$$(58)$$

The common aspect in these two examples is that the radicals may chemically act as oxygen radicals although they can physically not be accurately addressed as such. Corresponding considerations apply, in fact, to many other radicals whose characteristic chemical properties are not directly apparent and are hidden behind their dominant electronic resonance structure.

PROBLEMS OF CURRENT INTEREST (from the Radiation Chemistry point of view)

Research with oxygen centered radicals is a very active field in Radiation Chemistry, and many problems still await solution. Concerning ·OH radicals, for example, great interest is focused on its particular oxidizing action relative to simple one-electron oxidants. It touches, in particular, on the problem of selectivity and the mechanism of addition-elimination processes.

The field of $O_2^-/HO_2\cdot$ research is enormous and mainly related to biochemical and biological problems; any detailed analysis of the state of the art is not possible. Radiation and other chemists are particularly concerned with the reactivity, i.e., rate constants and mechanisms for the reactions initiated by these species in different environments.

A particularly hot field of study is the formation and reactivity of peroxyl radicals. It includes rate constant measurements of O_2 addition to radical centers (not only at carbon but also at other atoms), and the effect of substituents and molecular structures on the formation and the consecutive reactions of the peroxyl radicals. Only limited information is available at present on the redox properties of many peroxyl radicals.

Similar considerations hold for oxy radicals. Many of the problems concerning these species are connected to the peroxyl radicals, but also to peroxides and hydroperoxides. Further attention needs to be focused on possible differences of the reactivity of both peroxyl and oxy radicals in various solvents or environments.

Although noted as problems and open questions from a radiation chemical point of view, solutions and answers may nevertheless also be found by appropriate research in other fields. Radiation Chemistry is, however, a branch of science which is particularly suited for free radical research.

SUGGESTED READING

Asmus, K.-D., 1984, Pulse radiolysis methodology, in "Oxygen Radicals in Biological Systems," L Packer, ed., Academic Press, Orlando.

Asmus, K.-D., 1983, Sulfur-centered free radicals, in "Radio protectors and Anticarcinogens," O.F. Nygaard and M.G. Simic, eds., Academic Press, New York.

Asmus, K.-D., Bahnemann, D., Krischer, K., Lal, M., and Mönig, J., 1985, One-electron induced degradation of halogenated methanes and ethanes in oxygenated and anoxic aqueous solutions, *Life Chemistry Reports*, *3*, 1.

Asmus, K.-D., Möckel, H., and Henglein, A., 1973, Pulse radiolysis study on the site of $\cdot$OH radical attack on aliphatic alcohols in aqueous solution, *J. Phys. Chem.*, 77:1218.

Aust, S.D., Morehouse, L.A., and Thomas, G.E., 1985, Role of metals in oxygen radical reactions, *J. Free Radicals in Biol. and Med.*, 1:3.

Baxendale, J.H. and Busi, F., eds., 1982, "The Study of Fast Processes and Transient Species by Electron Pulse Radiolysis," NATO Advanced Study Institute, D. Reidel, Dordrecht.

Behar, D., Czapski, G., Rabani, J., Dorfman, L.M., and Schwarz, H.A., 1970, *J. Phys. Chem.*, *74*, 3209.

Bielski, B.H.J., 1984, Generation of superoxide radicals in aqueous and ethanolic solutions by vacuum UV photolysis, in "Methods in Enzymology," Vol. 105, pp. 81, L. Packer, ed., Academic Press, Orlando.

Bors, W., Saran, M., and Tait, D., eds., 1984, "Oxygen Radicals in Chemistry and Biology", W. de Gruyter, Berlin.

Bühler, R.E., Staehelin, J., and Hoigné, 1984, Ozone decomposition in water studied by pulse radiolysis.1. HO_2/O_2^- and HO_3/O_3^- as intermediates, *J. Phys. Chem.*, 88, 2560, 5450.

Czapski, G., 1984, Reactions of OH·, in "Methods in Enzymology," L. Packer, ed., Vol. 105, pp. 209, Academic Press, Orlando.

Forni, L.G., and Willson, R.L., 1984, Electron and hydrogen atom transfer reactions: determination of free radical redox potentials by pulse radiolysis, in "Methods in Enzymology," L., Packer, ed., Vol. 105, pp. 179.

Henglein, A., Schnabel, W., Wendenburg, J., 1976, "Einführung in die Strahlenchemie," Verlag Chemie, Weinheim.

Mönig, J., Göbl, M., and Asmus, K.-D., 1985, Free radical one-electron vs hydroxyl radical induced oxidation. Reaction of trichloromethylperoxyl radicals with simple and substituted aliphatic sulphides in aqueous solution, J. Chem. Soc., II, 647.

Mönig, J., Asmus, K.-D., Schaeffer, M., Slater, T.F., and Willson, R.L., 1983, electron transfer reactions of halothane peroxy free radicals, CF_3CHClO_2·: measurement of absolute rate constants by pulse radiolysis, J. Chem. Soc. Perkin Trans. II, 1133.

Mönig, J., Bahnemann, D., and Asmus, K.-D., 1983, One-electron reduction of CCl_4 in oxygenated aqueous solutions: a CCl_3O_2· free radical mediated formation of CL^- and CO_2, Chem. Biol. Interactions, 47, 15.

Packer, L., ed., 1984, Oxygen radicals in biological systems, in "Methods in Enzymology," Vol. 105, Academic Press, Orlando.

Packer, J.E., Slater, T.F., and Willson, R.L., 1978, Reactions of the carbon tetrachloride related peroxyl free radical CCl_3O_2· with amino acids: pulse radiolysis evidence, Life Sci., 23, 2617.

Roberts, J.L., Jr., Calderwood, T.S., and Sawyer, D.T., 1983, Oxygenation by superoxide ion of CCl_4, $FCCl_3$, $HCCl_3$, p,p' –DDT, and related trichloromethyl substrates ($RCCl_3$) in aprotic solvents, J. Amer. Chem. Soc., 105, 7691.

Schuchmann, M.N., and Sonntag, C.V., 1982, Hydroxyl radical induced oxidation of diethylether in oxygenated aqueous solution. A product and pulse radiolysis study, J. Phys. Chem., 86, 1995.

Schuchmann, M.N., Zegota, H., and Sonntag, C.V., 1985, Acetate peroxyl radical, ·$O_2CH_2COO^-$: a study on the γ-radiolysis and pulse radiolysis of acetate in oxygenated aqueous solution, Z. Naturforsch, 40b, 215.

Sehested, K., Holcman, J., and Hart, E.J., 1983, Rate constants and products of the reactions of e_{aq}, O_2^-· and H· with ozone in aqueous solution, J. Phys. Chem., 87, 1951.

Slater, T.F., 1982, in "Biological Reactive Intermediates II", Snyder et al., eds., pp. 575, Plenum Press, New York.

Spinks, J.W.T., and Woods, R.J., 1976, "Radiation Chemistry", J. Wiley & Sons, New York and Toronto.

Stein, G., 1968, "Radiation Chemistry of Aqueous Solutions", The Weizman Science Press of Israel in cooperation with Interscience London.

Swallow, A.J., 1973, "Radiation Chemistry", 1973, Longman, London.

Zegota, H., Schuchmann, M.N., and Sonntag, C.V., 1984, Cyclopentylperoxyl and Cyclohexylperoxyl radicals in aqueous solutions: a study by product analysis and pulse radiolysis, J. Phys. Chem., 88, 5589.

THE AUTOXIDATION OF POLYUNSATURATED LIPIDS

Ned A. Porter and Dennis G. Wujek

Paul M. Gross Chemical Laboratories
Duke University
Durham, North Carolina 27706

INTRODUCTION

The gradual accumulation of oxygen which accompanied the evolution of photosynthetic organisms approximately 1.5 billion years ago was most significant in that an appropriate environment had been provided for the genesis of aerobic organisms. Along with the evolution of respiratory organisms emerged a most interesting irony. The very molecular oxygen which had always been an agent of biological degradation became a vital commodity to the sustainment of life. That which had supported the lives of respiring organisms was also toxic to them, and it was only by virtue of the development of elaborate defense mechanisms that survival was made possible.

Today, a full understanding of the dual role of molecular oxygen has not yet been realized. Numerous studies have recently been carried out examining the beneficial and deleterious behavior of oxygen. The essential role of oxygen has been established in human vital processes such as liver detoxification (Jauhonen et al., 1985), collagen synthesis (Stryer, 1981; El-Far and Pimstone, 1985), and arterial cell wall repair (Hess and Manson, 1985). The injurious role of molecular oxygen, on the other hand, has been linked with biological events such as aging (Harman, 1985), heart disease (Chambers et al., 1985), and pulmonary failure (Kubow et al., 1985; Stuart et al., 1985). The attack of molecular oxygen on fatty acids and other lipids is thought to play an integral role in these processes; hence, lipid peroxidation has recently attracted considerable interest.

Molecular Oxygen

Atmospheric molecular oxygen is a ground state triplet. Since it contains two unpaired electrons, oxygen exhibits diradical behavior with univalent pathways of reduction favored over divalent pathways. In addition to acting as an oxidant, i.e., electron acceptor, molecular oxygen can also add to free radicals to yield peroxy radicals.

$$O_2 + e^- \rightarrow O_2^-\cdot \tag{1}$$

$$O_2 + R\cdot \rightarrow ROO\cdot \tag{2}$$

Both of these modes of reactivity are potentially destructive and capable of inducing deterioration of materials of biological importance.

Oxygen as Oxidant. It is generally thought that the reactivity of oxygen itself is rather poor. However, because its reduction to water proceeds by a series of single electron transfers (Williams, 1984), a number of highly reactive intermediates are generated in this process. It has been suggested that the intermediates in reduction such as hydrogen peroxide (H_2O_2), superoxide radical ($O_2^-\cdot$), and hydroxyl radical ($\cdot OH$) are responsible for the toxic character of oxygen. It should be emphasized that these same reduction products are also vital to virtually all living organisms. For instance, the same reagent ($O_2^-\cdot$) which can cause random free radical stress in a biological system may also act in a beneficial manner by attacking foreign bacteria. Of course, this scenario is dependent upon where and when the reagent is made and used. We shall focus our attention on the generation and subsequent harmful effects of these reduced intermediates, especially with regard to lipid damage.

Superoxide radical is the univalently reduced form of oxygen and is produced in most cells capable of reducing oxygen (Fridovich, 1984). Although a key intermediate in biological oxidation reactions (Curnutte, et al., 1974; Rosen and Freeman, 1985), $O_2^-\cdot$ has repeatedly been shown to be toxic and destructive to living cells when generated in enzymatic or photochemical fluxes and its mechanism of damaging action is not yet known (Fridovich, 1981).

Hydrogen peroxide, the most stable and least toxic of the aforementioned reduced intermediates, arises from either the dismutation of $O_2^-\cdot$ or from two consecutive monovalent reductions of oxygen. Both $O_2^-\cdot$ and H_2O_2 exhibit some toxic behavior towards cellular metabolism; however, they primarily appear to serve as precursors to the true oxidizing agent in many systems, the hydroxyl radical ($\cdot OH$).

$$O_2 \xrightarrow{e^-} O_2^-\cdot \xrightarrow[\text{or } e^-]{\text{dismutation}} H_2O_2 \xrightarrow{Fe^{2+}} \cdot OH \tag{3}$$

Electron transfer from Fe^{2+} to H_2O_2 serves as a major *in vivo* source of the extremely reactive $\cdot OH$. It is thought that $\cdot OH$, the most potent oxidant known (Fridovich, 1976), may be responsible for most oxidative lipid damage (Gutteridge, 1984; Bors et al., 1984).

The primary line of defense against the damaging influences of reduced forms of molecular oxygen in biological systems is probably provided by the enzymes superoxide dismutase (SOD), catalase, and peroxidase. SOD catalyzes the dismutation of $O_2^-\cdot$ (Fridovich, 1981); catalase and peroxidase scavenge H_2O_2 (Baldwin, et al., 1985; Frew and Jones, 1984). Their mode of action then is to effectively limit concentrations of $O_2^-\cdot$ and H_2O_2.

$$2O_2^-\cdot + 2H^+ \xrightarrow{\text{SOD}} O_2 + H_2O_2 \tag{4}$$

$$2H_2O_2 \xrightarrow[\text{catalase}]{\text{peroxidase or}} O_2 + H_2O \tag{5}$$

A secondary line of defense against *in vivo* oxidative damage is that of antioxidant activity (e.g., Vitamin E) which, in effect, serves to minimize peroxide and $\cdot OH$ concentrations in cell membranes (Williams, 1984; Bannister, 1984; Fridovich, 1979).

Oxygen as Reactant. Oxygen may act not only as an oxidant to initiate free radical reactions, but it can also act as a substrate for the propagation of these reactions. The spontaneous reaction of molecular oxygen with radicals is commonly referred to as autoxidation. Autoxidation is responsible for the deterioration of many manufactured plastics and rubber goods. Rancidity and spoilage of foodstuffs is a direct result of the autoxidation of fats which are most susceptible to air oxidation and present to a large extent in virtually all foods.

The autoxidative process is commonly represented as consisting of chain initiation, propagation, and termination steps:

Initiation: $\quad In\cdot + RH \rightarrow R\cdot + InH$ $\qquad\qquad\qquad\qquad\qquad$ (6)

Propagation: $\quad R\cdot + O_2 \rightarrow ROO\cdot$ $\qquad\qquad\qquad\qquad\qquad\qquad$ (7)

$$ROO\cdot + RH \xrightarrow{k_p} R\cdot + ROOH \qquad\qquad\qquad (8)$$

Termination: $\quad 2ROO\cdot \rightarrow [ROOOOR] \rightarrow \text{non--radical products} + O_2$ $\qquad$ (9)

The key event in initiation is the formation of $R\cdot$. There are many sources of radical species which may serve to abstract a hydrogen atom from RH. These include nitric oxide (NO), nitrogen dioxide (NO_2), and ozone, all of which are common environmental pollutants and have been shown to initiate autoxidation *via* hydrogen atom abstraction (Autor, 1984). Another method of generating $R\cdot$ is *via* thermal or photochemical induced homolytic scission:

$$RH \xrightarrow[\text{UV radiation}]{\text{heat or}} R\cdot + H\cdot \qquad\qquad\qquad\qquad\qquad (10)$$

In biological systems, potential chain-initiating radicals may be generated, for instance, *via* electron transfer processes:

$$M^{n+} + ROOH \rightarrow M^{(n+1)+} + {}^-OH + RO\cdot \qquad\qquad\qquad (11)$$

$$M^{n+} + O_2 \rightarrow M^{(n+1)+} + O_2^{-}\cdot \qquad\qquad\qquad\qquad\quad (12)$$

In the initial propagation step of autoxidation (Eq. 7), molecular oxygen adds to $R\cdot$. At partial oxygen pressures above 100 mm Hg, this addition approaches the diffusion controlled limit (*ca* 10^9 l/mol-s) (Ingold, 1969). This means that the major radical in solution is peroxy radical $ROO\cdot$ and not $R\cdot$. As a result, it is unlikely that any termination reactions involving $R\cdot$ will take place, provided that oxygen is present in sufficient concentration. In the second propagation reaction (Eq. 8), $ROO\cdot$ abstracts hydrogen atom from RH at rate k_p to generate more $R\cdot$ in what is the rate-determining step. For each hydroperoxide product formed, another radical $R\cdot$ is generated. This process could proceed *ad infinitum* save for termination reactions that interrupt propagation. One potential termination step shown is that of peroxy radical coupling to ultimately give oxygen and non-radical organic products (Eq. 9) (Howard, 1973; Russell, 1957). Another possible terminating route is that of radical disproportionation.

The rate constant k_p for hydrogen atom abstraction depends primarily on the activation energy of the reaction, i.e., the strength of the C-H bond being broken. Since ROO· is strongly resonance stabilized and a comparatively unreactive radical (Ingold, 1969), it is quite selective in abstraction from hydrocarbons and prefers the most weakly bound hydrogen atom. The selectivity of peroxy radicals has been demonstrated in experiments where hydrogen is abstracted in marked preference to deuterium, resulting in a large kinetic isotope effect (Howard et al., 1968):

$$k_p^{\alpha-H}/k_p^{\alpha-D} = 20.0$$

Another route of peroxy radical reaction is that of addition to a double bond to give the more stable β-peroxy alkyl radical. Addition will tend to be the favored route when the double bond is in conjugation with stabilizing groups such as vinyl, carbonyl, or aromatic.

$$ROO\cdot + CH_2{=}CHR \rightarrow ROOCH_2\dot{C}HR \tag{13}$$

Intramolecular peroxy radical addition reactions result in cyclic peroxides provided that orientation and ring size are favorable for such a reaction to occur:

Antioxidants

Since most organic compounds are subject to destruction by autoxidation, it is a matter of considerable importance to prevent such a reaction. Antioxidants (or inhibitors) accomplish this task by breaking kinetic chains or by preventing initiation. In the former category are included compounds such as aromatic amines and hindered phenols which both contain readily abstractable hydrogen atoms. Considering hydrogen atom removal from a phenol, the resultant phenoxy radical is sufficiently unreactive such that it will not enter into any processes to continue the chain. The observation that phenolic antioxidants react poorly with alkyl free radicals but most readily with peroxy radicals indicates that Eq. 14 represents their chain-breaking effect (Menzel, 1976).

$$\text{ROO·} + \text{ArOH} \xrightarrow{k_{inh}} \text{ROOH} + \text{ArO·} \qquad (14)$$

$$\text{ROO·} + \text{ArO·} \rightarrow \text{non–radical products} \qquad (15)$$

Perhaps the most well-known phenolic antioxidant is α-tocopherol, the major constituent of Vitamin E (Fig. 1). Ubiquitous in nature, α-tocopherol is one of the most powerful chain-breaking antioxidants known, reacting with peroxy radicals at rate k_{inh} on the order of 10^6 l/mol-s (Diplock, 1983; Burton et al., 1983; Burton, Hughes et al., 1983).

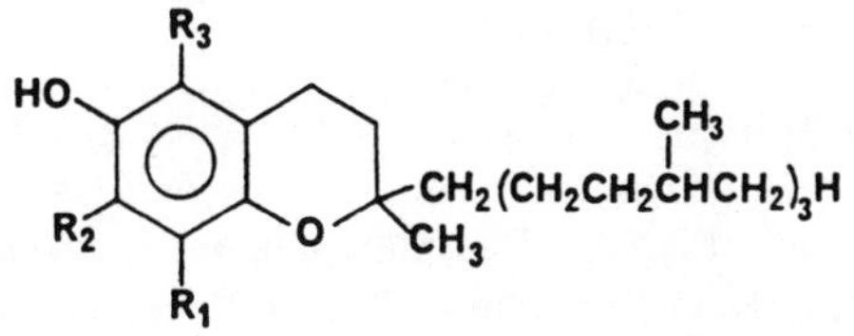

α- Tocopherol: $R_1 = R_2 = R_3 = CH_3$

β- Tocopherol: $R_1 = R_3 = CH_3$; $R_2 = H$

γ- Tocopherol: $R_1 = R_2 = CH_3$; $R_3 = H$

δ- Tocopherol: $R_1 = CH_3$; $R_2 = R_3 = H$

Figure 1. Vitamin E.

Table 1. Antioxidant activity of α-tocopherol and selected analogues.

$$\text{ROO·} + \text{ArOH} \xrightarrow{k_{inh}^{rel}} \text{ROOH} + \text{ArO·}$$

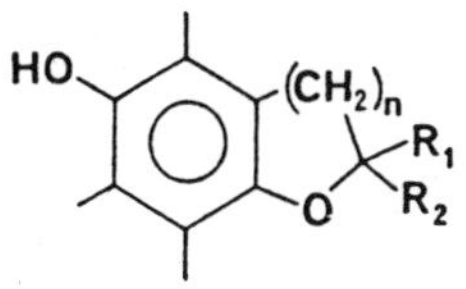

CMPD.	R_1	R_2	n	k_{inh}^{rel}
1	H	H	2	0.82
α-tocopherol, 2	$C_{16}H_{33}$	Me	2	1.00
3	Me	Me	2	1.16
4	Me	Me	1	1.66

Most interestingly, recent attempts have been successful at preparing analogues of α-tocopherol which match and, in some cases, even exceed the chain-breaking capabilities of the parent compound due to inherent stereoelectronic factors (Table 1) (Burton et al., 1983).

A disadvantage associated with many commonly used chain-breaking antioxidants is that they lose their inhibitory efficiency above *ca* 100° C due to homolytic decomposition of peroxide products formed in Eqs. 14 and 15. Hence, at elevated temperatures, a second class of inhibitors known as preventive antioxidants are used to block the formation of radicals which can initiate autoxidation. For instance, *ortho*-hydroxybenzophenones can be added to an organic compound and prevent autoxidation by preferentially absorbing any damaging radiation (Howard, 1973).

Polyunsaturated Fatty Acids (PUFAs)

Fatty acids are important components of biological membranes in that they serve as building blocks for major membrane constituents such as phospholipids, glycolipids, waxes, and triacylglycerols (Fig. 2). Phospholipids and glycolipids primarily provide a structural foundation for cell membranes. Triacylglycerols, because they are reduced and anhydrous, serve as stores for metabolic energy. Finally, waxes are generally metabolic endproducts and usually serve as protective chemical coatings. Fatty acids are mainly found in the form of one of these classes and exit freely in only trace quantities in biomembranes (Bohinski, 1979).

Structurally, most naturally occurring fatty acids are unbranched and contain an even number of carbon atoms (Table 2). They are typically 14 to 24 carbon atoms in length with 16- and 18-carbon fatty acids being the most common. The number of double bonds in a chain can vary from 0 to 6 with 0 to 2 being the most common. Virtually all natural fatty acids have *cis* alkene stereochemistry with multiple double bonds being separated by at least one methylene group.

A convenient system for designating structure has been described by Holman (1971). In this shorthand notation, the number of carbon atoms is followed by the number of double bonds in the chain (e.g., 18:1 or 18:2) and the positions of the double bonds are designated by a prefix denoting their distance from the carboxyl group (e.g., 9-18:1 or 9,12-18:2).

It is apparent from numerous reviews and articles concerning unsaturated fatty acids that these substances are necessary and important components of living tissue. This is exemplified in their ability to impart desirable properties upon the fluidity of membranes of which they are an integral part (Stryer, 1981). However, these essential unsaturated molecules can ultimately disrupt membrane stability by virtue of the fact that they are most susceptible to oxidative processes (Vanduijn et al., 1984). This is true because their methylene interrupted double bonds make them prime targets for peroxy radical attack. That is, as the degree of unsaturation increases, so too does the propensity for loss of hydrogen atom to give an increasingly stable, delocalized radical system which ultimately reacts further with molecular oxygen (Eq. 7). This is reflected by an increase in the rate constant k_H^{rel}, for hydrogen atom abstraction by peroxy radical ROO· from increasingly unsaturated fatty acids (Table 3) (Howard and Ingold, 1967).

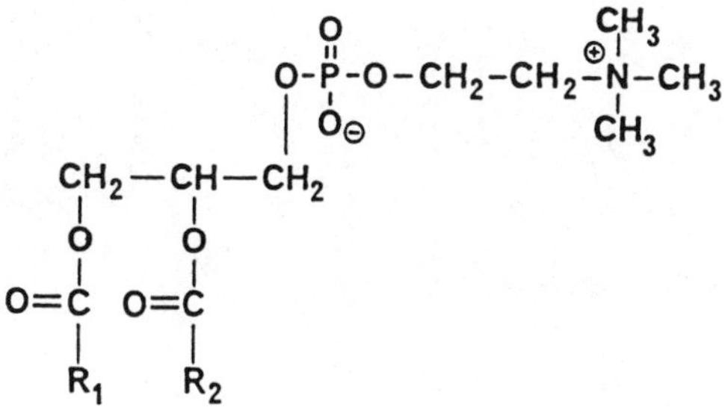

O−P−O−CH$_2$−CH$_2$−$\overset{\oplus}{N}$−CH$_3$

CH$_3$

CH$_3$

CH$_2$−CH−CH$_2$

O=C O=C

R$_1$ R$_2$

1,2-diacyl phosphatidylcholine
(a phospholipid)

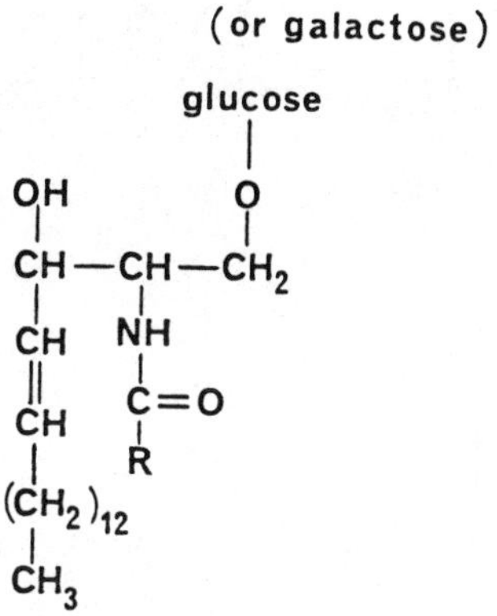

(or galactose)

glucose

OH

CH−CH−CH$_2$

CH   NH

CH   C=O

R

(CH$_2$)$_{12}$

CH$_3$

cerebroside
(a glycolipid)

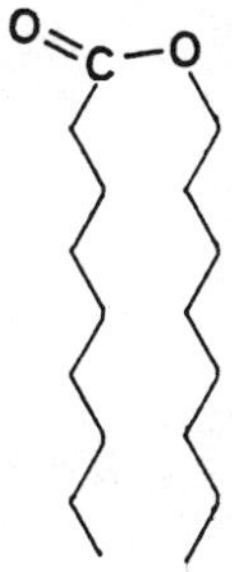

waxes

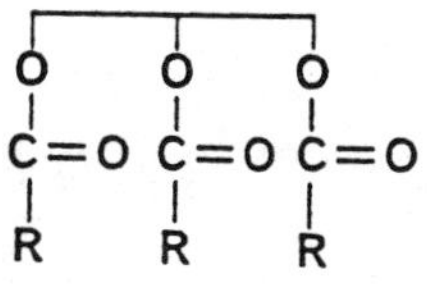

O O O

C=O C=O C=O

R   R   R

triacylglycerols

Figure 2. Some classes in which fatty acids serve as a major component.

Table 2. Some naturally occurring fatty acids.

Saturated: $CH_3-(CH_2)_n-CO_2H$

Structure	Common Name	Shorthand
n = 10	lauric	12:0
n = 12	myristic	14:0
n = 14	palmitic	16:0
n = 16	stearic	18:0
n = 18	arachidic	20:0
n = 20	behenic	22:0

Unsaturated:

Structure	Common Name	Shorthand
COOH	palmitoleic	c-9-16:1
COOH	oleic	c-9-18:1
COOH	linoleic	c,c-9,12-18:2
COOH	γ-linolenic	c,c,c-6,9,12-18:3
COOH	α-linolenic	c,c,c-9,12,15-18:3
COOH	arachidonic	all c-5,8,11,14-20:4

Table 3. Relative rates of hydrogen atom abstraction from some fatty acids by peroxy radical (Howard & Ingold, 1967).

$$R-H + R'OO\cdot \xrightarrow{\ k_H^{rel}\ } R\cdot + R'OOH$$

Fatty Acid (R-H)		k_H^{rel}	C-H bond strength[*] D(C-H), kcal/mole
stearic	~~~~~~~~~~COOH	<<1.0	95 (2°)
oleic	~~~~~=~~~~COOH	1.0	87 (allylic)
linoleic	~~=∨=~~~COOH	62.0	82 (bisallylic)

[*]Bond energy values determined from hydrogen atom abstraction by <u>t</u>-butoxy radical from corresponding fatty acid.[41]

Enzymatic Oxidation of PUFAs

It is known that virtually all plants have the ability to oxidize PUFAs to conjugated fatty acid hydroperoxides (FAHPs). Enzymes which catalyze this transformation are referred to as lipoxygenases. For example, maize lipoxygenase yields exclusively the 9-*d* -hydroperoxylinoleate from linoleic acid and the 5-*d* -hydroperoxyarachidonate from arachidonic acid (Scheme 1). In a similar manner, soybean lipoxygenase gives rise to the 13-*l*-hydroperoxylinoleate and 15-*l*-hydroperoxyarachidonate from linoleic and arachidonic acids respectively. The overall conversion to a conjugated diene hydroperoxide involves bisallylic hydrogen atom removal and subsequent addition of molecular oxygen. Both of these steps are thought to take place sterospecifically. It is known that addition of molecular oxygen is *anti* to hydrogen atom removal and the resultant diene sterochemistry is always *trans,cis* relative to the hydroperoxide functionality (Fig. 3) (Egmond et al., 1972). Further studies have established that hydrogen atom removal is rate-determining in the overall conversion.

Although lipoxygenase enzymes are universal in the plant kingdom, no biological function has ever been assigned directly to FAHPs resulting from lipoxygenase-mediated oxidation in plants. In sharp contrast to this observation is the fact that FAHPs resulting from lipoxygenase action in physiological systems are of considerable biological importance. For instance, in human platelets, lipoxygenase enzymes are able to stereospecifically oxygenate arachidonic acid to afford hydroperoxyeicosatetraenoic acids (HPETEs) which, following enzymatically controlled dehydration, lead to an important class of bioregulators known as leukotrienes (See LTA$_4$s, Scheme 2).

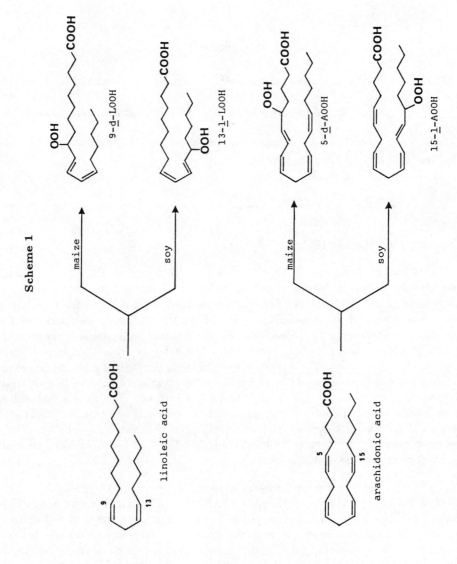

Scheme 1

64

Scheme 2

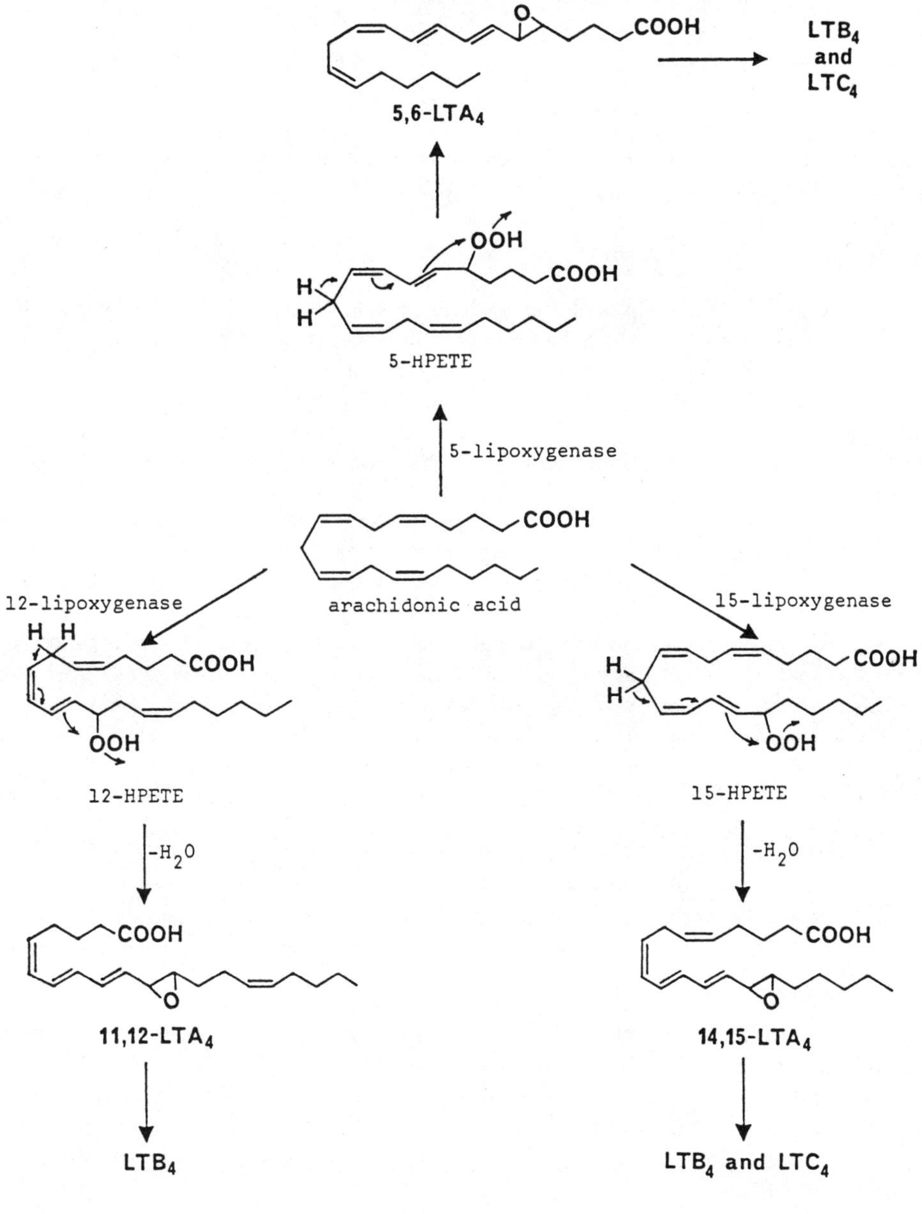

5,6-LTA$_4$ → LTB$_4$ and LTC$_4$

5-HPETE

5-lipoxygenase

arachidonic acid

12-lipoxygenase

15-lipoxygenase

12-HPETE

15-HPETE

$-H_2O$

$-H_2O$

11,12-LTA$_4$

14,15-LTA$_4$

LTB$_4$

LTB$_4$ and LTC$_4$

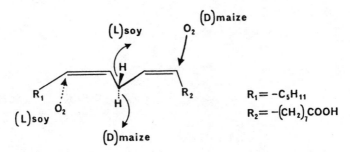

Figure 3. Stereospecific oxygenation of linoleic acid by either soy or maize lipoxygenase (Egmond et al., 1972).

Further reactions of the LTA_4s with nucleophiles such as cysteine and water lead to still more leukotrienes — all of which have unique and far-ranging biological effects. It is known, for example, that LTB_4 is a neutrophil activator causing chemotactic movement of leukocytes (Beckman et al., 1985); leukotrienes C_4, D_4 and E_4, all cysteinyl-containing leukotrienes and often referred to as slow-reacting substances of anaphylaxis (SRS-As), are potent bronchoconstrictors and have pro-inflammatory effects in humans (Samuelsson, 1983).

In addition to lipoxygenase, there are other enzymes which also convert arachidonic acid to biochemically important oxygenated compounds, all of which form an integral part of the Arachidonic Acid Cascade (Scheme 3) (Pike and Morton, 1985). In a process catalyzed by cyclooxygenase enzymes, arachidonic acid is stereospecifically oxygenated at the 11-position and the resultant 11-HPETE cyclized to give endoperoxide PGG_2 (Scheme 4). Chemical models mimicking this enzymatic conversion suggest that the reaction probably follows a free radical pathway (Nelson et al., 1982; Corey et al., 1984; Porter et al., 1984). This route includes two consecutive 5-exo radical cyclizations to form the 2,3-dioxabicyclo[2.2.1]heptane skeleton and subsequent entrapment of molecular oxygen and hydrogen atom to yield the final prostaglandin product.

Prostaglandins play significant roles in biological processes such as platelet aggregation, smooth muscle contraction, and inflammation. Suffice it to say that prostaglandins are a most important class of bioregulators which exhibit unique and varied effects (Samuelsson, 1983).

Also originating from PGH_2 are two other important classes of metabolites. One of these is a family of compounds known as thromboxanes. TXA_2 is a highly unstable bicyclic compound known to induce platelet aggregation and vascular smooth muscle contraction in humans. TXB_2, a hydrolysis product of TXA_2, is a much more stable derivative which was originally thought to be biologically inactive (Pace-Asciak and Granström, 1983).

Scheme 3

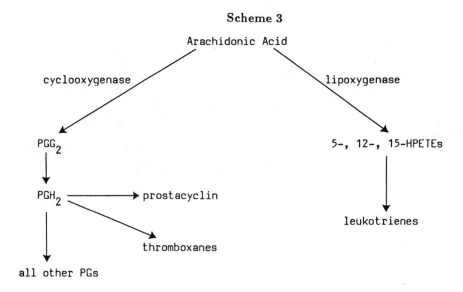

Not stored in the body but biosynthesized upon demand, PGG_2 is the precursor to PGH_2 which in turn is the prostaglandin from which all other PGs ultimately originate:

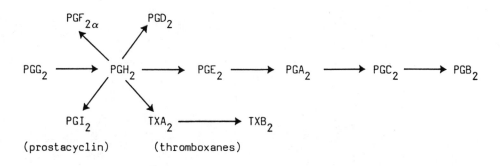

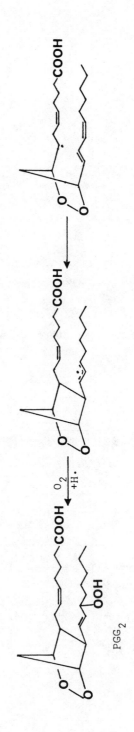

Scheme 4

arachidonic acid

cyclo-oxygenase

O_2

O_2
+H·

PGG$_2$

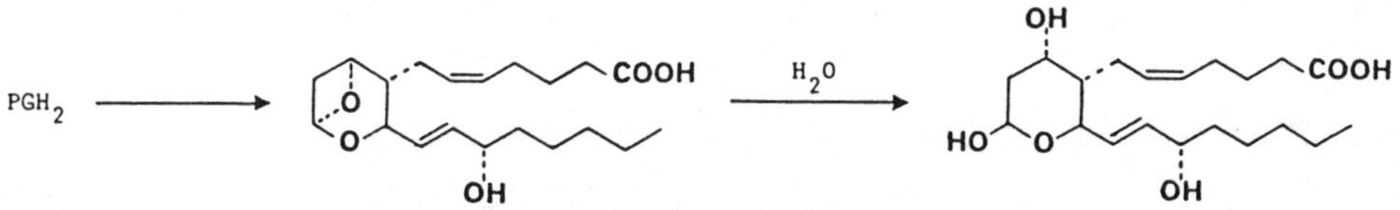

$$PGH_2 \longrightarrow \text{TXA}_2 \xrightarrow{H_2O} \text{TXB}_2$$

However, recent studies indicate that this compound also possesses some activity (Sirois et al., 1981; Peskar et al., 1984). Another type of metabolite finding its origin in PGH_2 is PGI_2, or prostacyclin. Prostacyclin is a known vasodilator and anti-aggregating compound and is also thought to form part of a hemostatic mechanism with TXA_2 for controlling tonus of blood vessels and platelet aggregation *in vivo* (Gryglewski et al., 1985).

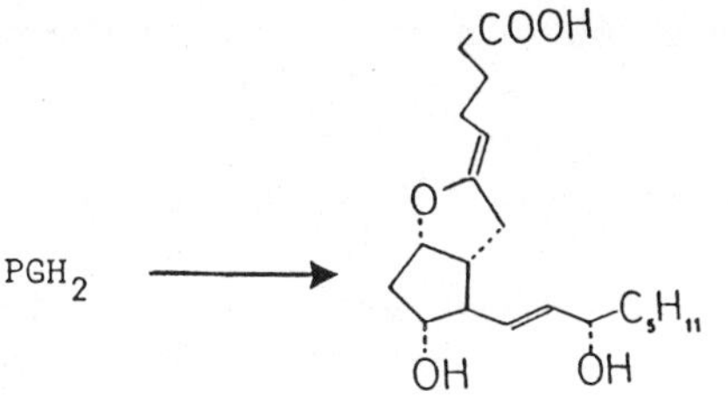

PGI_2 (prostacyclin)

In addition to the aforementioned chemical, nonenzymatic models for PCG_2 biosynthesis, there is other support for the intermediacy of free radicals in PUFA enzymatic oxidation. This includes ESR studies of the oxygenating enzymes and the observation that the biosynthesis of oxygenated intermediates can be reduced using free radical inhibitors (de Groot et al., 1975; Slappendel et al., 1981).

Summary

It is most apparent that molecular oxygen is a paradoxical element of nature. Intermediates resulting from the stepwise reduction of molecular oxygen are seen to be both beneficial and destructive. Similarly, we see that the products resulting from the enzymatic oxygenation of polyunsaturated fatty acids are also found to exhibit a dual behavior in nature. Next, we shall examine in far greater detail the mechanism by which molecular oxygen reacts with polyunsaturated fatty acids in a random, uncontrolled manner to afford some of the same oxygenated products whose diverse actions have been outlined herein.

UNIFIED MECHANISM FOR POLYUNSATURATED FATTY ACID (PUFA) AUTOXIDATION

Diene Fatty Acid Autoxidation

As discussed previously, most PUFAs have methylene interruped double bonds which are usually of *cis* geometry, as is the case for linoleic acid. It was noted that enzymatically, the *cis,cis* 1,4-diene system in linoleate was

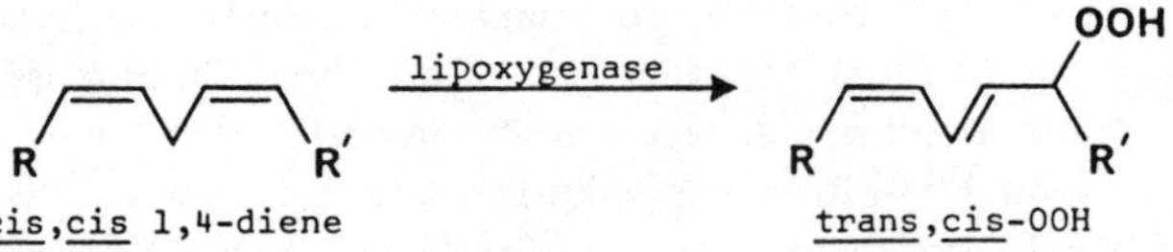

cis,cis 1,4-diene   trans,cis-OOH

converted to a *trans,cis* conjugated diene hydroperoxide with the double bond closest to the hydroperoxide functionality always possessing *trans* geometry. In the nonenzymatic oxidation of linoleate, however, the resultant products are seen to be more complex. Not just one but four major conjugated diene hydroperoxides 6-9 are isolated. Two products have *trans,cis* (t,c) and the other two products have *trans,trans* (t,t) diene stereochemistry (Chan and Levett, 1977). Two products result from the incorporation of molecular oxygen at the 9-position and the other two products result from oxygen addition at the 13-position. Together these products account for more than 97% of the oxygen consumed in linoleate autoxidation.

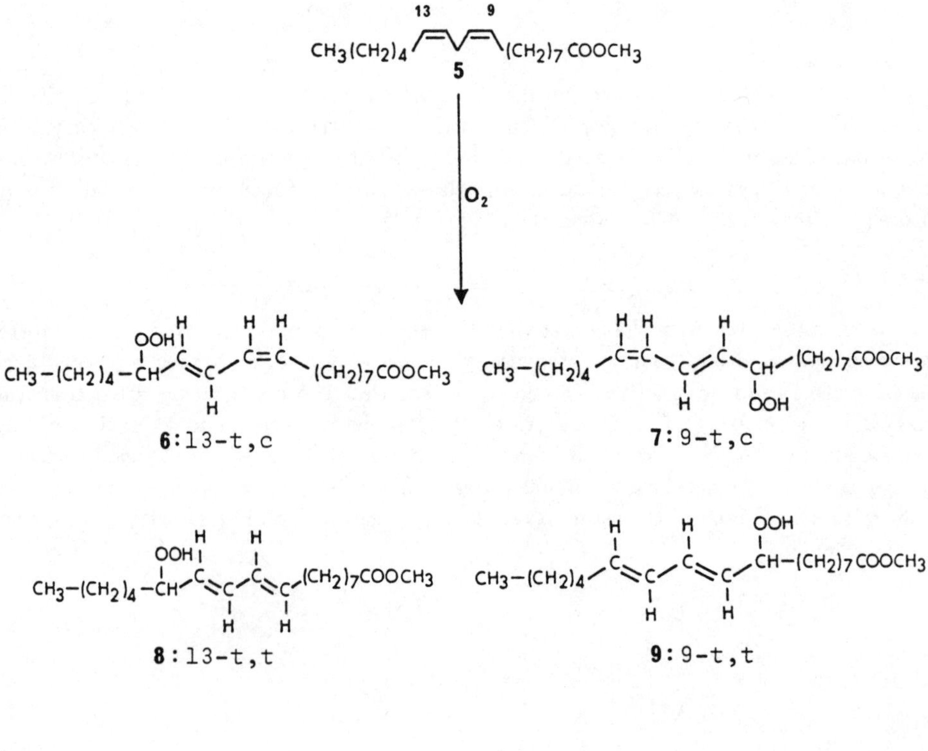

6:13-t,c   7:9-t,c

8:13-t,t   9:9-t,t

70

Most interestingly, the distribution of the four major diene hydroperoxide products varies greatly with the medium in which autoxidation occurs. For instance, it has been observed that the *trans,cis* to *trans,trans* product ratio [t,c]/[t,t] is less than 1.0 when linoleate is autoxidized in organic solvents such as benzene or acetonitrile at concentrations less than 1.0 M at room temperature (Porter et al., 1980; Yamamoto et al., 1982). However, autoxidation in the presence of an excellent H atom transfer agent such as α-tocopherol in benzene yields a trans,cis/trans,trans ratio of *ca* 70-75 (Smith, unpublished results). Before examining product stereochemistry in linoleate autoxidation, it is imperative that kinetic and product studies which formed a basis for the prior understanding of diene fatty acid autoxidation be cited here.

One important mechanistic consideration is the lack of dependence of product composition on oxygen concentration. It has been shown that the rate of linoleate autoxidation is independent of oxygen pressure between 100 mm and 760 mm Hg (Howard and Ingold, 1967). This implies that the primary chain-carrying radicals in fatty acid autoxidation are peroxy and not carbon radicals.

Another basic tenet for understanding PUFA autoxidation involves the high reactivity of bisallylic hydrogen atoms in 1,4-diene systems. As mentioned previously, carbon-hydrogen bonds at a bisallylic position are relatively weak and are therefore most susceptible to H atom abstraction by a linoleate peroxy radical (Table 3). Hydrogen atom abstraction from linoleate results in the formation of a pentadienyl system with a "W" configuration (Thomas and Pryor, 1980).

Other possible configurations for pentadienyl radicals resulting from a 1,4-diene system include the "Z" and "U" conformers. Formation of these pentadienyl systems would be highly unlikely due to the additional strain involved in having three or four cisoid interactions for the "Z" and "U" systems respectively.

71

It should be noted, however, that in a study of minor linoleate autoxidation products (Roberts, 1982), a small amount (<1%) of oxidation product was proposed to originate from the "Z" linoleate conformation:

R$_1$ $\xrightarrow{-H\cdot}$ R$_1$ "Z" $\xrightarrow{O_2}$ HOO R$_1$ 9-cis,cis

R$_2$ $\xrightarrow{-H\cdot}$ R$_2$ "Z" $\xrightarrow{O_2}$ HOO R$_2'$ 13-cis,cis

Another concept integral to the understanding of product distribution in linoleate autoxidation is the reversible addition of oxygen to pentadienyl systems. This was demonstrated when linoleate hydroperoxides were found to rearrange in a manner such that the hydroperoxide functionality was relocated and the geometry of a double bond was changed. For instance, it was shown in an oxygen labelling study that a single unlabelled *trans,cis* hydroperoxylinoleate could equilibrate under $^{36}O_2$ atmosphere to labelled *trans,cis* and *trans,trans* hydroperoxylinoleates (Chan et al., 1979).

R' OOH trans,cis ⇌ R' OO· ⇌

R +O$_2$ ⇌ ·OO R' ⇌ HOO R' trans,trans

A free radical mechanism was invoked to account for the arrangement. Implicit in the mechanism was the suggestion that the reaction of molecular oxygen with pentadienyl radicals was reversible. A most important consequence of this study was that pentadienyl carbon radicals could isomerize only *via* the addition and elimination of molecular oxygen to and from the pentadienyl system. That there is no direct isomerization of pentadienyl carbon radicals has been supported by several recent EPR studies (Griller et al., 1979; Davies et al., 1981; Bascetta et al., 1983).

With the advent of high pressure liquid chromatographic methods for the separation of linoleate hydroperoxides, a more thorough understanding of linoleate autoxidation was realized (Porter et al., 1980). Most notably, it was found that the trans,cis/trans,trans product ratio varied with temperature and concentration but was independent of time. The product ratio was observed to increase with increasing linoleate concentration at all temperatures and decrease with increasing temperature at constant linoleate concentrations. The fact that [t,c]/[t,t] was invariant over the time course of an experiment (less than 5% oxidation) implied that the product distribution was kinetically controlled. That is, no significant amount of hydroperoxylinoleate equilibrated once it had formed. This was a particularly important point since it had previously been demonstrated that linoleate hydroperoxides underwent facile thermal equilibration at conditions as mild as heating at 40° C for 16 hours (Chan et al., 1979). In the same 1980 linoleate study by Porter (Porter et al., 1980), a mechanism was proposed to account for the formation of the four major linoleate hydroperoxides and invoked to explain the observed dependence of [t,c]/[t,t] on temperature and fatty acid concentration (Scheme 5). As is apparent from this scheme, the basis for understanding product distribution in linoleate autoxidation involves the reversible addition of oxygen to intermediate pentadienyl carbon radicals.

In the proposed mechanism, H atom abstraction from fatty acid **5** leads to "W" radical **10**. Molecular oxygen can then add to either end of the delocalized system to afford peroxy radicals **14** and **17**. Considering oxygen addition to one terminus of pentadienyl system **10**, we see that the resultant peroxy radical **17** can actually exist in two different conformations, **17a** sand **17b** (Scheme 6). β-Scission of oxygen (where the C-O bond undergoing fragmentation is perpendicular to the diene π-system) from conformer **17a** leads back to the initial pentadienyl system **10**. β-Scission, however, from conformer **17b** leads to an isomerized carbon radical **13** which has *trans,cis* dienyl stereochemistry. Therefore, it is through this reversible oxygen addition process that one or both of the double bonds in the initial pentadienyl system **10** may be isomerized Hydrogen atom abstraction by peroxy radical **17a/17b** from the medium leads to *trans,cis* hydroperoxide poduct whereas H atom abstraction by peroxy radical **16a/16b** leads to *trans,trans* hydroperoxide product. The crucial competition then which determines the trans,cis/trans,trans product ratio is β-scission of peroxy radical **17a/17b** to ultimately give *trans,trans* product or H atom abstraction by radical **17a/17b** to give *trans,cis* hydroperoxide product. At high concentrations of linoleate, the *trans,cis* peroxy radicals are trapped before they have time to undergo β-scission. At low concentrations of linoleate, the unimolecular process of β-scission (k_β) is more competitive with the bimolecular process of H atom abstraction (k_p) so that *trans,trans* products predominate. The terms α and 1-α are simply inserted to account for the partitioning of isomerized carbon radical **13** between the *trans,cis* and *trans,trans* peroxy radical manifolds. Because of symmetry, a similar argument can be made for carbon radical **11** and peroxy radicals **14** and **15**.

Scheme 5

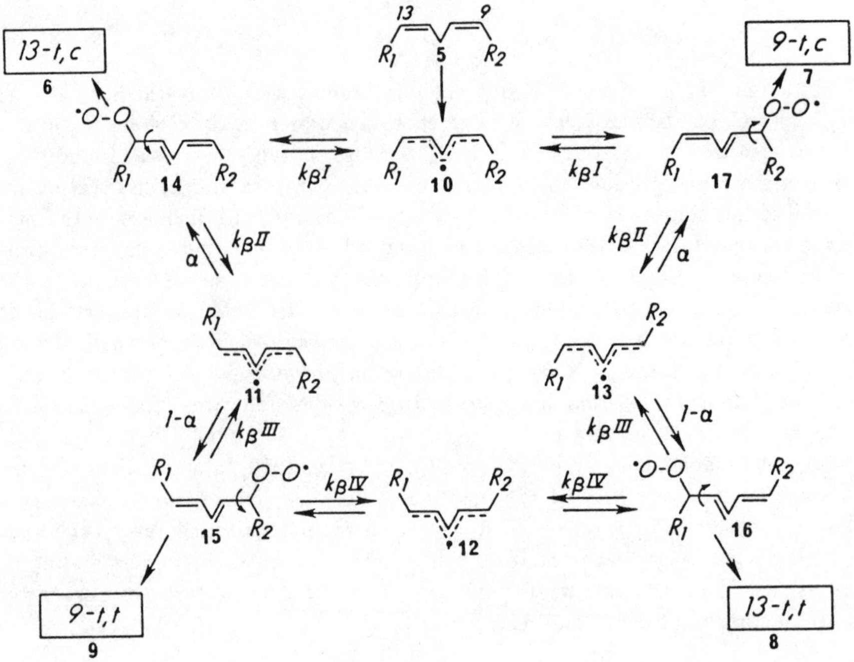

Scheme 6

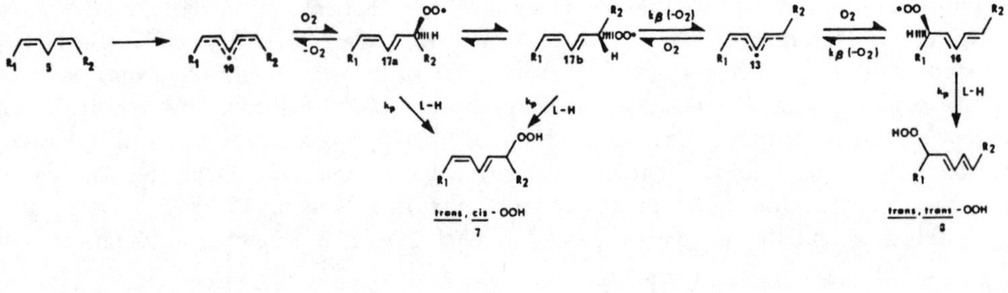

Applying steady-state assumptions in the analysis of Scheme 5, a kinetic expression was derived for relating the trans,cis/trans,trans product ratio to k_p^L, α, and all the k_β's. Assuming a common k_β (i.e., $k_\beta{}^{II} = k_\beta{}^{III}$), Eq. 16 was derived where k_p^L is the known rate of propagation for linoleate autoxidation ($k_p^L = 62$ M^{-1}s^{-1}) (Howard and Ingold, 1967). Equation 16, then, predicted that the product ratio $(\mathbf{6+7})/(\mathbf{8+9})$ should be linearly dependent on linoleate concentration L–H with the slope $= k_p^L/k_\beta(1-\alpha)$ and intercept $= \alpha/(1-\alpha)$.

$$\frac{\text{trans,cis}}{\text{trans,trans}} = \frac{\mathbf{6+7}}{\mathbf{8+9}} = \frac{k_p^L[\text{L–H}]}{k_\beta(1-\alpha)} + \frac{\alpha}{1-\alpha} \tag{16}$$

Analysis in this manner led to the calculated values of $\alpha = 0.14$ and $k_\beta = 144$ s^{-1}. It should be reemphasized that these calculations were made with the assumption of a common k_β.

It should also be noted that studies were carried out where various H atom donating compounds were used as cosubstrates for the autoxidation of linoleate (Porter et al., 1981). In the absence of an excellent H atom donor, *trans,trans* products were seen to be predominant.

In the presence of excellent H atom donors such as 1,4-cyclohexadiene or 9,10-dihydroanthracene, the bimolecular process of H atom abstraction was more competitive and *trans,cis* products were predominant. A refined kinetic expression was presented to describe $[t,c]/[t,t]$ as it related to the total hydrogen atom donating ability of the medium, KP:

$$\frac{[t,c]}{[t,t]} = \frac{KP}{k_\beta(1-\alpha)} + \frac{\alpha}{1-\alpha} \tag{17}$$

$$\text{where } KP \equiv \sum_{i=1}^{n} k_{p_i}[R_i-H]$$

Using this expression, rate constants for hydrogen atom transfer (k_p) from various cosubstrates to linoleate peroxy radicals were calculated. Shown in Fig. 4 is a sample calculation of rate constant k_p as carried out for 1,4-cyclohexadiene. Once again, note that these calculations of k_p were made with a common k_β value ($k_\beta = 144$ s^{-1}) and a value of α calculated from the autoxidation of linoleate alone at 30 °C.

The concept of using this new parameter, KP, is not new to free radical kinetics. The idea of carrying out competition kinetics by varying the H atom availability of the medium has been a cornerstone of many mechanistic studies (Beckwith and Ingold, 1980). The definition of KP here is justified because the linoleate autoxidation product

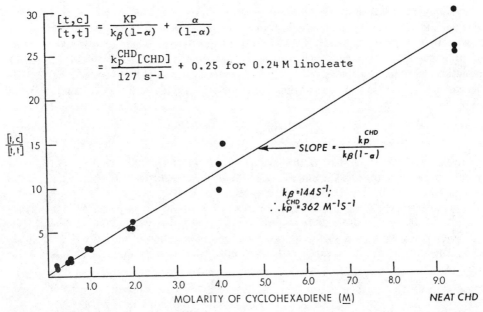

Figure 4. Plot of [t,c]/[t,t] versus concentration of 1,4-cyclohexadiene/benzene for linoleate autoxidation (Porter et al., 1981).

distribution depends directly on KP. The extrapolation of product studies of linoleate autoxidation to systems of unknown composition could potentially give information about the KP of that system.

Triene and Tetraene Fatty Acid Autoxidation

Autoxidation of PUFAs with three or more double bonds deserves brief comment here. In addition to the β-scission and H atom abstraction processes, a third reaction pathway available to triene or tetraene fatty acids is the unimolecular process of peroxy radical cyclization. Cyclization, for instance, of peroxy radical **18** at rate k_c affords radical **19** which can undergo further reactions to yield products such as monocyclic peroxides, bicyclic peroxides, and epoxy alcohols (O'Connor et al., 1984; Zuraw, 1985).

Utilizing the unimolecular β-scission pathway as a "free radical clock" (Griller and Ingold, 1980) for timing H atom abstraction from donors such as 1,4-cyclohexadiene, the rate of peroxy radical cyclization was calculated to be $8 \times 10^2 \, s^{-1}$. The kinetics of triene and tetraene fatty acid autoxidation have been discussed in greater detail elsewhere (Porter et al., 1981).

REFERENCES

Autor, A.P., 1984, in "The Biology and Chemistry of Active Oxygen," p. 140, J.V. Bannister and W.H. Bannister, eds., Elsevier, New York.

Baldwin, D.A., Marques, H.M., and Pratt, J.M., 1985, *FEBS Lett.*, *183 (2)*, p. 309.

Bannister, W.H., 1984, in "The Biology and Chemistry of Active Oxygen," pp. 217-219, J.V. Bannister and W.H. Bannister, eds., Elsevier, New York.

Bascetta, E., Gunstone, F.D., and Walton, J.C., 1983, *J. Chem. Soc., Perkin Trans. II*, 603.

Beckman, J.K., Gray, J.C., Brash, A.R., Lukens, J.N., and Oates, J.A., 1985, *Lipids*, 20(6):357.

Beckwith, A.L., and Ingold, K.U., 1980, "Rearrangements in Ground and Excited States," 1, 161, P. De Mayo, ed., Academic Press, New York.

Bohinski, R.C., 1979, "Modern Concepts in Biochemistry," 3rd ed., pp. 316-344, Allyn and Bacon, Boston.

Bors, W., Saran, M., and Tait, D., eds., 1984, "Oxygen Radicals in Chemistry and Biology," pp. 137-145 and pp. 285-297, de Gruyter, Berlin.

Burton, G.W., Cheeseman, K.H., Doba, T., Ingold, K.U., and Slater, T.F., 1983, in *Ciba Found. Symp., 101 (Biol. Vitam. E)*, pp. 4-18.

Burton, G.W., Hughes, L., and Ingold, K.U., 1983, *J. Am. Chem. Soc.*, 105:5950.

Chambers, D.E., Parks, D.A., Patterson, G., Roy, R., McCord, J.M., Yoshida, S., Parmley, L.F., and Downey, J.M., 1985, *J. Mol. Cell. Cardiol.*, 17(2):145.

Chan, H.W.-S., and Levett, G., 1977, *Lipids*, 12:99.

Chan, H.W.-S., Levett, G., and Matthew, J.A., 1979, *Chem. Phys. Lipids*, 24:245.

Corey, E.J., Shimoji, K., and Shih, C., 1984, *J. Am. Chem. Soc.*, 106:6425.

Curnutte, J.T., Shitten, D.M., and Babior, B.M., 1974, *N. Engl. J. Med.*, 290:593.

Davies, A.G., Griller, D., Ingold, K.U., Lindsay, D.A., and Walton, J.C., 1981, *J. Chem. Soc., Perkin Trans. II*, 633.

de Groot, J.J.M.C., Veldink, G.A., Vliegenthart, J.F.G., Boldingh, J., Wever, R., and van Gelder, B.F., 1975, *Biochim. Biophys. Acta*, 377:71.

Diplock, A.T., 1983, in *Ciba Found. Symp., 101 (Biol. Vitam. E)*, pp. 45-55.

Dogra, S.C., Khanduja, K.L., Gupta, M.P., and Sharma, R.R., 1985, *Ind. J. Med. Res.*, 81:520.

Egmond, M.R., Vliegenthart, J.F.G., and Boldingh, J., 1972, *Biochim. Biophys. Res. Comm.*, 48:1055.

El-Far, M.A., and Pimstone, N.R., 1985, *Cell. Biochem. Funct.*, 3(2):115.

Fridovich, I., 1984, in "The Biology and Chemistry of Active Oxygen," pp. 128-138, J.V. Bannister and W.H. Bannister, eds., Elsevier, New York.

Fridovich, I., 1981, in "Oxygen and Oxy-radicals in Chemistry and Biology," pp. 197-204, M.A.J. Rodgers and E.L. Powers, eds., Academic Press, New York.

Fridovich, I., 1979, in "Oxygen Free Radicals and Tissue Damage," *Ciba Found. Symp. 65*, pp. 1-4, Excerpta Medica, Amsterdam.

Fridovich, I., 1976, in "Free Radicals in Biology," 1:249, W.A. Pryor, ed., Academic Press, New York.

Frew, J.E., and Jones, P., 1984, *Adv. Inorg. Bioinorg. Mech.*, 3:175.

Griller, D., and Ingold, K.U., 1980, *Acc. Chem. Res.*, 13:317.

Griller, D., Ingold, K.U., and Walton, J.C., 1979, *J. Am. Chem. Soc.*, 101:758.

Gryglewski, R.J., Szczeklik, A., and McGill, J.C., eds., 1985, "Prostacyclin Clinical Trials," Raven Press, New York.

Gutteridge, J.M.C., 1984, *FEBS Lett.*, 172(2):245.

Harman, D., 1985, *Age (Omaha, NE)*, 7(4):111.

Hess, M.L., and Manson, N.H., 1985, *J. Mol. Cell. Cardiol.*, 16(11):969.

Holman, R.T., ed., 1971, "Progress in the Chemistry of Fats and Other Lipids," V. 9, Pergamon Press, Oxford.

Howard, J.A., 1973, in "Free Radicals," V. 2, pp. 3-62, J.K. Kochi, ed., Wiley (Interscience), New York.

Howard, J.A., Ingold, K.U., and Symonds, M., 1968, *Can. J. Chem.*, 46:1017.

Howard, J.A., and Ingold, K.U., 1967, *Can. J. Chem.*, 45:793.

Ingold, K.U., 1969, *Acc. Chem. Res.*, 2:1.

Jauhonen, V.P., Baraona, E., Lieber, C.S., and Hassinen, I.E., 1985, *Alcohol*, 2(1):163.

Kubow, S., Bray, T.M., and Janzen, E.G., 1985, *Biochem. Pharmacol.*, 34(7):1117.

Kurisaki, E., 1985, *J. Toxicol. Sci.*, 10(1):29.

Menzel, D.B., 1976, in "Free Radicals in Biology," V. 2, p. 192, W.A. Pryor, ed., Academic Press, New York.

Nelson, N.A., Kelley, R.C., and Johnson, R.A., August 16, 1982, *Chem. & Eng. News*, p. 30.

O'Connor, D.E., Mihelich, E.D., and Coleman, M.C., 1984, *J. Am. Chem. Soc.*, 106:3577.

Pace-Asciak, C., Granström, E., eds., 1983, "Prostaglandins and Related Substances," pp. 50-52, Elsevier, New York.

Peskar, B.A., Zimmerman, I., and Ulmer, W.T., 1984, *Klin. Wochenschr.*, 62(7):315.

Pike, J.E., and Morton, Jr., D.R., eds., 1985, "Advances in Prostaglandin, Thromboxane, and Leukotriene Research," V. 14, pp. 156-157, Raven Press, New York.

Porter, N.A., Zuraw, P.J., and Sullivan, J.A., 1984, *Tet. Lett.*, 25:807.

Porter, N.A., Lehman, L.S., Weber, B.A., and Smith, K.J., 1981, *J. Am. Chem. Soc.*, 103:6447.

Porter, N.A., Weber, B.A., Weenen, H., and Khan, J.A., 1980, *J. Am. Chem. Soc.*, 102:5597.

Pryor, W.A., Prier, D.G., Lightsey, J.W., and Church, D.F., 1980, in "Autoxidation in Food and Biological Systems," pp. 1-16, M.G. Simic and M. Karel, eds., Plenum Press, New York.

Roberts, D.H., 1982, *Ph.D. Dissertation*, Duke University, pp. 16, 155.

Rosen, G.M., and Freeman, B.A., 1985, *Proc. Natl. Acad. Sci. USA*, 81(23):7269.

Russell, G.A., 1957, *J. Am. Chem. Soc.*, 79:3871.

Samuelsson, B., 1983, *Science*, 220:568.

Samuelsson, B., 1983, *Angew, Chem. Int. Engl. Ed.*, 22:816.

Sharabani, M., Plotkin, B., and Aviram, I., 1984, *Cell. Mol. Biol.*, 30(4):329.

Sinha, B.K., Trush, M.A., and Kalyanaraman, B., 1985, *Biochem. Pharmacol.*, 34(11):2036.

Sirois, P., Borgeat, P., and Jeanson, A., 1981, *J. Pharm. Pharmacol.*, 33:466.

Slappendel, S., Veldink, G.A., Vliegenthart, J.F.G., Aasa, R., and Malström, B.G., 1981, *Biochim. Biophys. Acta*, 667:77.

Stryer, L., 1981, "Biochemistry," 2nd ed., pp. 186-187, 205-209, W.H. Freeman and Co., New York.

Stuart, R.S., Baumgartner, W.A., Borkon, A.M., Bulkley, G.B., Brawn, J.D., De La Monte, S.M., Hutchins, G.M., and Reitz, B.A., 1985, *Transplant. Proc.*, 17(1, Bk. 2):1454.

Thomas, M.J., and Pryor, W.A., 1980, *Lipids*, 15:544.

Vanduijn, G., Verkleij, A.J., and Dekruijf, B., 1984, *Biochem.*, 23:4969.

Warson, M.A., and Lands, W.E.M., 1983, *Br. Med. Bull.*, 39(3):277.

Williams, R.J.P., 1984, in "The Biology and Chemistry of Active Oxygen," pp. 1-15, J.V. Bannister and W.H. Bannister, eds., Elsevier, New York.

Yamamoto, Y., Niki, E., and Kamiya, Y., 1982, *Lipids*, 17:870.

Zuraw, P.J., 1985, *Ph.D. Dissertation*, Duke University.

ORGANIC PEROXY FREE RADICALS AND THE ROLE OF SUPEROXIDE DISMUTASE AND ANTIOXIDANTS

R.L. Willson

Department of Biochemistry
Brunel University
Uxbridge, Middlesex, U.K.

When a biological system is exposed to free radicals, clearly a large number of different molecules will be affected. Some of these molecules may be important, for example DNA or perhaps a key protein whose normal rate of synthesis is comparatively slow. The repercussions may then be serious. If other molecules are affected this may be less important: destruction of a small amount of glucose, for example, is unlikely to effect a biological system per se. Of course, there does remain the possibility that stable or free radical products derived from such harmless substances may themselves be toxic. It must not be forgotten that when a free radical reacts with an organic molecule, another free radical is usually formed. The reactions of this radical must then be taken into account.

In recent years, radiation studies have proved invaluable for the characterisation of free radical reactions. It is still widely believed that when systems are exposed to radiation a host of more or less random reactions take place and that the free radicals involved are somewhat different from those that might be formed by conventional chemistry. It cannot be overstressed that in most instances, indeed in all the experiments referred to here, *radiation-induced free radicals are of thermal energy and follow the same laws of chemistry as other free radicals.*

A variety of biological systems have now been described where the changes occurring on exposure to ionising radiation can be explained by the involvement of free radical reactions.

It has long been known that bacteria and other cells are sensitive to ionising radiation and that this sensitivity is greater when oxygen is present. More recently, it has also been shown that various oxidising (sometimes described as "electron affinic") compounds (X) can also sensitise cells to radiation-induced killing. Two of such compounds, the drugs metronidazole and misondisole, have since been tested clinically as an adjunct

81

in the radiotherapy of cancer. The development of these drugs stemmed directly from simple chemical models involving free radicals and subsequent studies with bean roots, bacteria, mammalian cells systems and animals. According to the "repair-fixation" model of free radical damage, if a carbon-centred radical is formed on a vital molecule by hydrogen abstraction, this may be subsequently repaired by hydrogen transfer from a thiol compound such as glutathione.

Alternatively, however, if oxygen or another oxidising molecule is present, the free radical centre may be further oxidised, with the result that the damage may become permanent or "fixed". (Unfortunately, the use of the term "fixed" in this context has led to some confusion since the term is sometimes used to mean "repair" as in "fixing" a car. This is the exact reverse of what is meant in this free radical context.)

$$RH + OH\cdot \rightarrow R\cdot + H_2O$$

$$R\cdot + GSH \rightarrow RH + GS\cdot \qquad \text{(REPAIR)}$$

$$R\cdot + O_2 \rightarrow RO_2\cdot \qquad \text{(FIXATION/POSSIBLE DAMAGE)}$$

or $\quad R\cdot + O_2 \rightarrow R(-H) + O_2^-\cdot + H^+$

or $\quad R\cdot + X \rightarrow RX\cdot$

or $\quad R\cdot + X \rightarrow R(-H) + X\cdot^- + H^+$

Whether such reactions do occur in a particular biological system, will depend on the concentrations of local thiols, such as GSH, oxygen, and X, as well as and the nature of R·, which will clearly influence the associated reaction rate constants.

Although in even the simplest of biological systems, local concentrations of particular molecules are difficult to assess, absolute reaction rate constants are known with considerable degree of accuracy in many instances. Much of this information has been obtained using the technique of pulse radiolysis, details of which have been described elsewhere. The technique, the high energy analogue of flash photolysis, continues to provide much information concerning free radical mechanisms. For example, it is now known that whereas aliphatic carbon-centred radicals do not readily accept electrons from electron donors, oxygen-, nitrogen-, and sulphur centred radicals often do. Indeed, model experiments have shown that thiol compounds can provide a link between carbon-centred radicals and electron rather than hydrogen donors.

$R\cdot + GSH \rightarrow RH + GS\cdot$ $\qquad$ rapid hydrogen transfer

$R\cdot + NADH \rightarrow RH + NAD\cdot$ $\qquad$ slow hydrogen or electron transfer

$GS\cdot + NADH \rightarrow GSH + NAD\cdot$ $\qquad$ rapid electron transfer

The hydroxyl free radical has been extensively studied by pulse radiolysis, and the rate of its reaction with a range of compounds have been well documented.

Unfortunately, peroxy radicals in general are less reactive than the hydroxyl radical, and their reactions are not so conveniently studied using this technique. The halogen-substituted peroxy radicals such as those related to carbon tetrachloride are an exception. The rate constants of reaction of the trichlormethyl peroxy radical with a variety of antioxidants and biological molecules have now been measured, and have enabled experiments to be designed whereby the absorption spectrum of important lipid soluble free radicals, such as those of β-carotene and vitamin E, can be obtained.

Stationary state radiation experiments where the biological effects of longer periods of radiation (minutes rather than microseconds) are assessed have shown that the trichlormethyl peroxy free radicals and those derived from the reaction of OH· with several biological molecules in the presence of oxygen, can inactivate certain proteins and damage biological reducing agents under conditions where the superoxide radical, O_2^-·, does not.

Organic peroxy radicals derived from thymine and phenylalanine in particular have also been shown to be able to inactivate alcohol dehydrogenase and lactate dehydrogenase as well as the alpha-1-protease inhibitor. The presence of several antioxidants, including glutathione, NADH, Trolox C (a water soluble derivative of vitamin E) and superoxide dismutase can afford protection. Apo-SOD, prepared by dialysing the enzyme against cyanide, cannot afford protection unless traces of copper ion are also present.

$$RO_2· + NADH \rightarrow RO_2H + NAD· \quad \text{rapid}$$

$$O_2^-· + NADH \rightarrow \text{products} \quad \text{slow}$$

$$RO_2· + \text{Enzyme or virus} \rightarrow \text{damage (inactivation)}$$

$$RO_2· + \text{antioxidant} \rightarrow \text{products (protection)}$$

Interestingly, although uric acid has been reported to be an effective antioxidant in several lipid containing systems, there is a possibility that in aqueous protein systems it might accentuate free radical damage: the superoxide radical does not inactivate alcohol dehydrogenase in these experiments unless urate is also present.

In summary, a large number of free radical reactions with biological molecules have now been characterised *in vitro* using radiation methods. Whether such reactions are relevant *in vivo* not only to radiation-induced damage, but also to damage induced by what has been variously described as "decompartmentalised", "uncontrolled", "free", or "low molecular weight chelatable iron", awaits to be determined.

REFERENCES

Willson, R.L., 1977, Iron, zinc, free radicals and oxygen in tissue disorders and cancer control, Ciba Foundation, Symp. 51. Iron Metabolism, *Elsevier/Excerpta Medica North Holland*, pp. 331.

Gee, C.A., Kittridge, K.J., Willson, R.L., 1985, Peroxy free radicals, enzymes, and radiation damage: sensitization by oxygen and protection by superoxide dismutase and antioxidants, *Br. J. Radiol.*, 58:251.

Willson, R.L., 1985, Organic peroxy free radicals as ultimate agents in oxygen toxicity, in "Oxidative Stress," H. Sies, ed., pp. 41-72, Academic Press, London.

Willson, R.L., Dunster, C.A., Forni, L.G., Gee, C.A., and Kittridge, C.A., 1985, Organic free radicals and proteins in biochemical injury: electron or hydrogen transfer reactions?, *Phil. Trans. R. Soc. Lond.*, B311:545.

AN INTRODUCTION TO ELECTRON SPIN RESONANCE AND ITS
APPLICATION TO THE STUDY OF FREE RADICAL METABOLITES

Ronald P. Mason

Laboratory of Molecular Biophysics
National Institute of Environmental Health Sciences
Research Triangle Park, North Carolina

INTRODUCTION

Electron spin resonance (ESR) is one of a class of magnetic resonance experiments. Nuclear magnetic resonance (NMR) is a better known member of this class, but these preliminary comments will apply equally to electron spin resonance and nuclear magnetic resonance. From the names alone it is clear that electrons and nuclei, which are fundamental particles, are involved, and that the magnetic interactions of these particles are of predominant importance. The principal ideas behind these magnetic resonance spectroscopies depend on two characteristics of these fundamental particles. The first is that they are charged, and the second is that they have angular momentum or spin. Angular momentum is the motion of a body about a center, like the rotation of the earth. Angular momentum is a vector quantity which has both magnitude and direction. Another characteristic of fundamental particles is that quantum theory governs their energy, angular momentum and other physical properties. The main result of quantum mechanics is that the physical properties are discrete and not continuous as in classical mechanics. Quantum theory demands that the component of an electron's angular momentum in any given direction, call it J_z, be quantized,

$$J_z = m \frac{h}{2\pi} \text{ (erg sec)} \tag{1}$$

where J_z is the component of angular momentum in the direction z, where h is Planck's constant and m is the angular momentum quantum number. For an electron, m is limited to the discrete values $\pm 1/2$. You can think of this as (clockwise and counter clockwise) rotation or spin. The nuclear spin quantum number is usually denoted by I, and for ^1H, $I = \pm 1/2$ and for ^{13}C, $I = \pm 1/2$. Note that for ^{16}O and ^{12}C, which are the predominant isotopes of these atoms, $I = 0$, and, therefore, there is no ^{16}O NMR or ^{12}C NMR.

In summary, fundamental particles such as electrons have an intrinsic angular momentum which corresponds closely to the angular momentum associated with the spinning of a classical object about its axis. The major difference is that the spin of fundamental particles is quantized.

Now a charged particle with non-zero angular momentum or spin will, from freshman physics, generate a magnetic field. The magnetic field, called a magnetic moment, is proportional to the spin or angular momentum.

$$\mu_z = \gamma J_z \tag{2}$$

The proportionality constant γ is called the gyromagnetic ratio. Now, if a sample of free electrons, such as can exist in liquid ammonia, is placed in a laboratory magnetic field, H, where the direction of H can be taken to be in the aforementioned z direction, then quantum states with different values of the quantum number m will have different energies. The energies of the two states corresponding to $m = \pm 1/2$ can be written as the product of the magnetic field of the electron times the laboratory magnetic field.

$$E_\pm = -\mu_z \cdot H_z = -\gamma J_z \cdot H_z = -\gamma(m\frac{h}{2\pi})H_z = \pm \frac{1}{2}\gamma\frac{h}{2\pi}H_z \tag{3}$$

These two energy levels correspond to the magnetic moment or field of the electron being aligned either with or against the field of the laboratory magnet (Fig. 1).

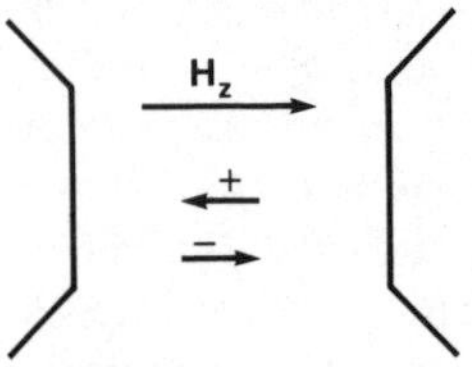

Figure 1. Possible orientations of an electron's magnetic moment in a laboratory magnet.

The populations in the two magnetic energy levels are related by the Boltzman factor

$$n_+/n_- = e^{-(E_+ - E_-)/kT} = e^{-\Delta E/kT} = e^{-(h/2\pi)\gamma H_z/kT} \tag{4}$$

where n_+ is the population in the upper level and n_- the population in the lower level, k is Boltzman's constant and T is the absolute temperature. The magnetic interactions are very weak, such that the magnetic interaction energy is much less than the thermal energy, kT, $\Delta E = (h/2\pi)\gamma H_z << kT$. Because $e^{-x} \approx 1 - x$ when $x << 1$ and $x = (h/2\pi)\gamma H_z/kT \approx 10^{-7}$, the Boltzman exponential can be approximated by

$$n_+/n_- \approx 1 - (h/2\pi)\gamma H_z/kT < 1 \tag{5}$$

We see that there is a slightly greater population in the lower energy level. Now that we have energy levels, we can induce transitions between them by exposing the sample to microwave electromagnetic radiation of a particular frequency υ (Fig. 2).

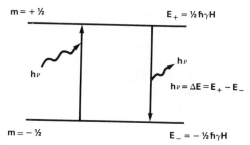

Figure 2. Radiation-linked transitions between the two energy levels of an electron in a magnetic field.

Note that the transition probability of an induced transition is the same in either direction. Only because of the population difference due to the Boltzman distribution will more electromagnetic energy or photons be absorbed than emitted. Therefore, the sample will absorb a net amount of electromagnetic energy, which is detected as the electron spin resonance signal.

$$\Delta E = h\upsilon = \Delta E = \frac{h}{2\pi}\gamma H_z \tag{6}$$

For typical magnetic fields H_z of 3000-10,000 G, υ is in the range of GHz or microwaves.

In summary, the energy of a photon of frequency υ is $h\upsilon$, and when it equals $(h/2\pi)\gamma H_z$, the energy difference between the magnetic states, the condition for energy absorption is fulfilled. This is the fundamental resonance condition as is commonly expressed in nuclear magnetic resonance.

It is customary in electron spin resonance to define

$$\frac{h}{2\pi}\gamma = g\beta \tag{7}$$

where β is the Bohr magneton, which is the classical gyromagnetic ratio, γ, of an electron. The g factor is a dimensionless constant that is characteristic of the molecular orbit of the electron.

For a free electron, g = 2.0023. In an organic free radical where the electron is delocalized over many atoms or nearly free, g is a little greater than 2.0023 and is characteristic of a class of free radicals. Iron usually also has unpaired electrons, and iron-containing enzymes such as cytochrome P-450 have ESR spectra characterized by their characteristic g factors.

For PMR, proton magnetic resonance, exactly the same formulas are used.

$$\Delta E = h\upsilon = \frac{h}{2\pi}\gamma H_z = g\beta_N H_z \tag{8}$$

The fundamental particle is the proton, but as for the electron, the spin quantum number is $\pm 1/2$. For typical laboratory magnetic fields, v is in the range of radio frequency. β_N is again the classical gyromagnetic ratio of the proton or the nuclear magneton. g is characteristic of the chemical environment of the proton. That is, methyl group protons can be distinguished from methylene protons. In order to "confuse" the issue, a different notation is used in NMR.

$$\Delta E = g\beta_N H = g\beta_N(1 - \sigma_N)H_z \qquad (9)$$

In this case, instead of defining a new g for each chemically different proton, the chemical shift notation is used where $g(1 - \sigma_H)$ replaces the chemical effect on g, and different classes of protons have different chemical shifts σ_H.

Now for an example of the use of g-values in the identification of a free radical. Mason et al. (1977) detected the electron spin resonance spectrum of a free radical metabolite of the azo dye sulfonazo III in anaerobic rat hepatic microsomal incubations containing the azo dye and NADPH. NADPH is the ultimate source of the extra electron in this biological system (Fig. 3).

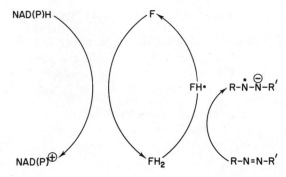

Figure 3. The proposed mechanism of azo radical anion formation. The exact scheme of electron donation by the microsomal flavoenzyme NADPH-cytochrome P-450 reductase (F) to azo compounds is unknown. (From Mason et al., 1977, with permission.)

This spectrum is characterized by a partially resolved 17-line hyperfine pattern and a g-value of the center line equal to 2.0034 (Fig. 4). g-Value measurements are analogous to chemical shift measurements in nuclear magnetic resonance and are used to characterize the structure of free radicals. DPPH (2,2-diphenyl-1-picrylhydrazyl free radical) is a g-value standard analogous to TMS (tetramethylsilane) in nuclear magnetic resonance. The 8 lines up-field and 8 lines down-field of the center line indicate that the unpaired electron is delocalized onto at least one of the aromatic rings and probably onto both azo groups. These lines are due to the interaction of the odd electron with the nuclei of the azo compound, and more will be said about them later.

Table I. g-Values of the sulfonazo III radical metabolite and other nitrogen containing organic radicals.

Radical	Isotropic g–Value[a]
Sulfonazo III radical metabolite	2.0034
Azobenzene anion radicals (t-butyl substituted)	2.0036
2,2-Diphenyl-1-picrylhydrazl (DPPH)	2.00354
Diphenylnitroxide	2.0055
Nitrobenzene anion radical	2.0049
Wurster's blue cation (N,N,N',N'-tetramethyl- p-phenylenediamine cation radical)	2.003051

[a]Not corrected for second order shifts.

The g-value of the sulfonazo III metabolite is indistinguishable from the g-values of the azobenzene anion free radical or the chemically related DPPH (Table I). This result suggests that the sulfonazo III radical metabolite is, in fact, an azo anion free radical. Unfortunately, the g-values of most free radicals are within 2% of the free electron value of 2.0023, and a coincidental equality of g-values must be considered. Therefore, a second group of organic radicals is included in Table I. These radicals have unpaired spin density localized on a nitrogen atom but are not chemically related to the azo anion free radical. Note that Wurster's blue cation has a g-value similar to, but significantly different from, that of the sulfonazo III radical metabolite.

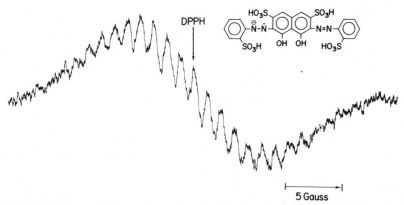

Figure 4. The ESR spectrum of the sulfonazo III radical anion metabolite (most probably a penta anion) seen in an anaerobic hepatic microsomal incubation containing an NADPH-generating system. The g value of 2,2-diphenyl-1-picrylhydrazl (DPPH) is indicated by the arrow. The assignmnet of the unpaired electron to one of the azo groups is made for the sake of convenience. (From Mason et al., 1978, with permission.)

The large number of nitrogens and hydrogens in sulfonazo III allows over 2000 lines due to nuclear hyperfine interactions and is too poorly resolved to analyze, so let's consider a single nitrogen interacting with an electron. This interaction, like that of the laboratory magnetic field with the electron, is a magnetic interaction due to the nuclear magnetic moments.

One of the interactions between the electronic and nuclear magnetic moments is the same as the classical interaction between two bar magnets. It depends upon the orientation of the radical, however, and as the radical tumbles rapidly and randomly in solution, it averages to zero. Another interaction, which is explained by relativistic quantum mechanics but has no classical analog, is isotropic, and thus remains nonzero even for radical tumbling in solution. It is of the form

$$A' \; \mu_e \cdot \mu_N \tag{10}$$

where μ_e is the electron magnetic moment, μ_N the nuclear magnetic moment, and A' is a constant proportional to the probability of finding the unpaired electron at the position of the nucleus.

If we consider a radical containing a nucleus of spin 1 such as ^{14}N, then the component of nuclear spin angular momentum in a specific direction for such a nucleus can be only ± 1 or 0 times $h/2\pi$ ($m_I = \pm 1$ or 0). The nuclear magnetic moment is also proportional to the nuclear spin angular momentum. Thus, the possible interaction energies of the electron and nuclear magnetic moments are $\pm A/2$ and 0, where A is proportional to A'. For usual magnetic field strengths, A is much less than $g\beta H$.

The approximate energy level scheme for such a system is shown in Fig. 5. If H_z is held constant and the frequency υ increased, the usual absorption spectrum results. Due to the method of detection, electron spin resonance spectra are recorded as a first derivative of the absorption. The selection rules for the absorbtion of a microwave photon require that the electron spin quantum number change from $-1/2$ to $+1/2$. The nuclear spin quantum number can not change simultaneously.

In practice, in both ESR and NMR for reasons of experimental convenience, the frequency is usually kept fixed at a value υ_o, and the strength of the external field H varies. This gives a magnetic field-dependent energy level diagram (Fig. 6).

The spectrum, which now appears as a function of the field strength H, is the same as that which would be observed if H were kept fixed and υ varied. Note, however, that the transitions, as labeled by the m_I value, occur in reversed order with increasing independent variable. The hyperfine interaction energy, A, is customarily reported in units of field strength (gauss) rather than in units of energy or frequency. Note that transitions which occur at higher magnetic fields have energy levels which are closer together at a given magnetic field. So increasing magnetic field is, in a sense, going to transitions of lower frequency or energy.

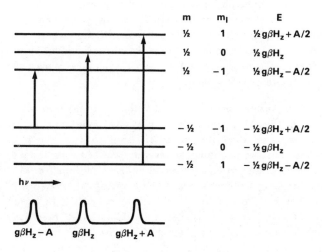

m	m_I	E
½	1	$\frac{1}{2}g\beta H_z + A/2$
½	0	$\frac{1}{2}g\beta H_z$
½	−1	$\frac{1}{2}g\beta H_z - A/2$
−½	−1	$-\frac{1}{2}g\beta H_z + A/2$
−½	0	$-\frac{1}{2}g\beta H_z$
−½	1	$-\frac{1}{2}g\beta H_z - A/2$

$h\nu \longrightarrow$

$g\beta H_z - A \qquad g\beta H_z \qquad g\beta H_z + A$

Figure 5. Energy levels for a spin-one nucleus, such as nitrogen and transitions for fixed field and variable frequency (not to scale).

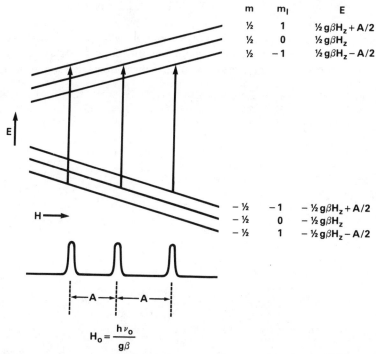

m	m_I	E
½	1	$\frac{1}{2}g\beta H_z + A/2$
½	0	$\frac{1}{2}g\beta H_z$
½	−1	$\frac{1}{2}g\beta H_z - A/2$
−½	−1	$-\frac{1}{2}g\beta H_z + A/2$
−½	0	$-\frac{1}{2}g\beta H_z$
−½	1	$-\frac{1}{2}g\beta H_z - A/2$

E

$H \longrightarrow$

$\longleftarrow A \longrightarrow \longleftarrow A \longrightarrow$

$$H_o = \frac{h\nu_o}{g\beta}$$

Figure 6. Energy levels for a spin-one nucleus, such as nitrogen and transitions for fixed frequency and variable field (not to scale).

As noted before, the hyperfine interaction energy is proportional to the probability of finding the unpaired electron at the position of the nucleus. If there are several magnetic nuclei in the radical, as there are with sulfonazo III, the hyperfine interactions add to each other, and each is proportional to the probability of finding an unpaired electron at that nucleus. The ESR spectrum of a free radical thus gives a map of the probability distribution of the unpaired electron spin.

An example of the use of nuclear hyperfine interactions is in identification of *cis* and *trans* isomers of the AF-2 anion radical by Kalyanaraman et al. (1979, 1981). Enzyme-catalyzed *cis-trans* isomerization of *cis*-AF-2 to *trans*-AF-2 by a variety of enzymes which reduce nitro compounds has recently received considerable attention (Fig. 7). *Cis*-AF-2 was used as an antibacterial food additive in Japan until it was found to be carcinogenic, mutagenic and cytotoxic. These biological activities of AF-2 and of other nitro compounds are thought to be initiated by enzymatic nitroreduction. Several investigators have observed that the isomerization of *cis*-AF-2 precedes nitroreduction. Tatsumi et al. (1976) have proposed that this enzymatic *cis*-trans isomerization is a direct consequence of enzymatic nitroreduction (Fig. 7). The nitroreductases, which are inhibited by oxygen, transfer a single electron to nitro substrates to give their respective anion free radicals. The carbon-carbon double bond linking the furan rings of AF-2 would be weakened by anion radical formation, because the additional electron is in an antibonding molecular orbital. Upon formation, the *cis*-AF-2 anion free radical was proposed to isomerize quickly to the *trans*-AF-2 anion, which could then be oxidized to form *trans*-AF-2. Tatsumi et al. (1976) showed that a purified nitroreductase, buttermilk xanthine oxidase, would catalyze the isomerization of *cis*-AF-2 to *trans*-AF-2 as well as the reverse reaction, and that the nitro group was necessary for this activity, but they provided no other evidence in support of their enzymatic isomerization mechanism.

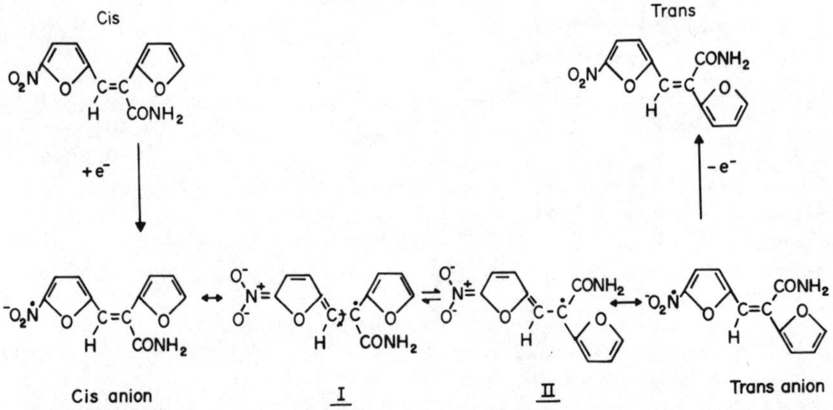

Figure 7. Postulated mechanism for the *cis-trans* isomerization of AF-2 by nitro-reducing enzymes. (From Kalyanaraman et al., 1981, with permission.)

ESR studies by Kalyanaraman et al. (1979, 1981) have clarified the nature of the relationship between the isomerization and the nitro reduction of AF-2. The ESR spectrum of an anaerobic microsomal incubation containing AF-2 and an NADPH-generating system provides direct evidence of free radical formation (Fig. 8). In the presence of air, nitroaromatic anion free radicals undergo rapid air oxidation to form superoxide anion, and the nitroaromatic anion free radicals are not detected. In addition, identical spectra were obtained with either *cis*- or *trans*-AF-2. These ESR spectra of the free radical intermediates did not vary with time even though visible spectroscopy shows a rapid and nearly complete conversion of *cis*-AF-2 to *trans*-AF-2 under the same conditions (Kalyanaraman et al., 1979).

The less abundant species represented by the two outer lines of the composite spectrum shows no detectable g-value shift from the predominant species, implying that it is also a nitrofuran anion free radical. The spectrum of the less abundant species is largely obscured and, for the purposes of computer simulation, we have assumed that the ESR parameters of the two free radicals differ only in their nitrogen hyperfine coupling constant. This assumption was made because the nitrogen coupling of nitrofuran anions is highly variable, and a small increase in this coupling is all that is necessary to accommodate the greater width of the less intense spectra. The best computer simulation assumed a 23:77 ratio for the concentration of the two nitrofuran anion free radicals (Kalyanaraman et al., 1981).

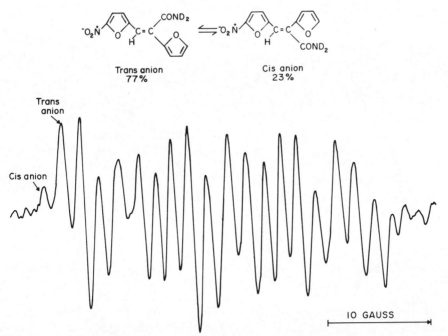

Figure 8. The first derivative ESR spectrum of a mixture of *trans*- and *cis*-AF-2 anion free radicals observed on anaerobic incubation of 2.0 mM *cis*-AF-2 (or *trans*-AF-2) with an NADPH generating system and hepatic microsomal protein in buffer made with D_2O. (From Mason, 1982, with permission.)

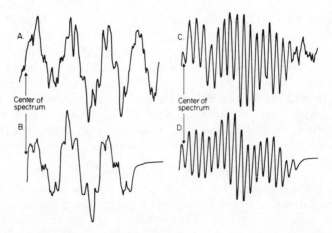

Center of
spectrum

Center of
spectrum

Figure 9. A. The ESR spectrum of the first four lines to the right of the center of the spectrum in Fig. 8. B. A computer-simulated ESR spectrum using all of the hyperfine couplings except $a_{NH_2}^H = 0.65$ G, which has been replaced by $a_{ND_2}^D$. Only those lines with $M_I = 0$ for $a_{NO_2}^N$ have been used in the simulation, therefore the fourth line in Fig. 9A which arises from $M_I = -1$ is not simulated. C. The ESR spectrum of the same region shown in Fig. 9A. All experimental conditions are the same except the buffer was made with H_2O instead of D_2O. Under these conditions, the unresolved fourth line seen clearly in Fig. 9A is undermodulated and is not distinguishable from the noise. D. A computer-simulated ESR spectrum using all the hyperfine couplings with $M_I = 0$ for $a_{NO_2}^N$. (From Kalyanaraman et al., 1981, with permission.)

The center of the spectrum can be resolved and provides another test of the accuracy of the hyperfine coupling constants (Fig. 9). This part of the spectrum corresponds to $M_I = 0$ for $a_{NO_2}^N$.

With the assumption that the hyperfine couplings of *cis*- and *trans*-AF-2 are the same except for $a_{NO_2}^N$ and that the reduction in the number of resolved lines upon substituting ^2H_2O for H_2O in the buffer is due to the exchange of the two hydrogen atoms of the amide group, the simulations in ^2H_2O-buffer and H_2O-buffer were obtained (Fig. 9).

The assignments of the two $a_{NO_2}^N$ hyperfine splitting constants are made primarily on the basis of molecular orbital calculations, which show that *cis*-$a_{NO_2}^N$ is greater than *trans*-$a_{NO_2}^N$. The experimental difference (*cis*-$a_{NO_2}^N$ minus *trans*-$a_{NO_2}^N$) = 1.7 G is in good agreement with the theoretical 1.2 G (Kalyanaraman et al., 1981). The assignment of the larger $a_{NO_2}^N$ to the *cis* conformer is also consistent with the lower expected relative abundance of this radical, because the crowding of the two furan moieties in the *cis* isomer is much more severe.

Although the thermodynamic ratio of *cis* to *trans*, or the conformation free energy, is similar to both the anion radical and the parent AF-2, the barrier to rotation is much lower for the anion radical. Whereas hours at elevated temperatures are required to obtain equilibrium of the parent AF-2 isomers, the half-time for isomerization of the *cis* anion radical has been estimated at only 17 ms from the effect of oxygen on the rate of isomerization (Kalyanaraman et al., 1979). This rate of isomerization has been confirmed by pulse radiolysis (Clarke, et al., 1984).

For a variety of reasons, the possibility of free radical metabolism has not received much attention in the past, although Michaelis of the Michaelis-Menten equation was interested in free radical metabolites and their importance in biochemistry in the 1930's. One reason for the later development of this area is that most biochemicals are not aromatic, and therefore are not easily metabolized through free radical intermediates. Glutathione and abscorbate are rather rare exceptions to this rule. In contrast, almost all synthetic aromatic organic chemicals and many inorganic chemicals such as bisulfite are metabolized to reactive free radicals by at least one enzyme.

Past investigations of free radical metabolites using electron spin resonance have been reviewed (Mason, 1982) including the spin trapping (Mason, 1984a), the fast flow (Yamazaki, 1977), and the steady-state *in situ* (Mason, 1984b) techniques. Wertz and Bolton (1972) is recommended as an introductory text for electron spin resonance, but it is not recommended that one teach oneself the theory and practice of electron spin resonance, especially the spin trapping technique.

REFERENCES

Clarke, E.D., Wardma, P., Wilson, I., and Tatsumi, K., 1984, The mechanism of the free-radical-induced chain isomerization of 2-(2-furyl)-3-(5-nitro-2-furyl)acrylamide, *J. Chem. Soc. Perkin Trans.*, II:1155.

Kalyanaraman, B., Mason, R.P., Rowlett, R., and Kispert, L.D., 1981, An electron spin resonance investigation and molecular orbital calculation of the anion radical intermediate in the enzymatic *cis-trans* isomerization of furylfuramide, a nitrofuran derivative of ethylene, *Biochim. Biophys. Acta*, 660:102.

Kalyanaraman, B., Perez-Reyes, E., Mason, R.P., Peterson, F.J., and Holtzman, J.L., 1979, Electron spin resonance evidence for a free radical intermediate in the *cis-trans* isomerization of furylfuramide by oxygen-sensitive nitroreductases, *Mol. Pharmacol.*, 16:1059.

Mason, R.P., 1982, Free-radical intermediates in the metabolism of toxic chemicals, in "Free Radicals in Biology, Vol. V," W.A. Pryor, ed., Academic Press, New York.

Mason, R.P., 1984a, Spin trapping free radical metabolites of toxic chemicals, in "Spin Labeling in Pharmacology", J.L. Holtzman, ed., Academic Press, New York.

Mason, R.P., 1984b, Assay of *in situ* radicals by electron spin resonance, in "Methods in Enzymology, Vol. 105, Oxygen Radicals in Biological Systems," L. Packer, ed., Academic Press, New York.

Mason, R.P., Peterson, F.J., and Holtzman, J.L., 1977, The formation of an azo anion free radical metabolite during the microsomal azo reduction of sulfonazo III, *Biochem. Biophys. Res. Commun.*, 75:532.

Mason, R.P., Peterson, F.S., and Holtzman, J.L., 1978, Inhibition of azoreductase by oxygen. The role of the azo anion free radical metabolite in the reduction of oxygen to superoxide, *Mol. Pharmacol.*, 14:665.

Wertz, J.E., and Bolton, J.R., 1972, "Electron Spin Resonance, Elementary Theory and Practical Applications," McGraw-Hill Book Company, New York.

Yamazaki, I., 1977, Free radicals in enzyme-substrate reactions, in "Free Radicals in Biology, Vol. III," W.A. Pryor, ed., Academic Press, New York.

DETERMINATION OF DISSOLVED OXYGEN IN PHOTOSYNTHETIC SYSTEMS AND PROTECTIVE STRATEGIES AGAINST ITS TOXIC EFFECTS IN CYANOBACTERIA

Shimshon Belkin and Lester Packer

Membrane Bioenergetics Group
Applied Science Division
Lawrence Berkeley Laboratory
Berkeley, California 94720

INTRODUCTION

In the study of the inhibitory effects of oxygen, photosynthetic organisms occupy a unique position. Unlike other biological systems, in which the cells are ordinarily exposed only to ambient O_2 levels and often far below, the oxygenic photosynthetic cell produces its own oxygen from within. The O_2 concentrations to which photosynthetic organisms are exposed, are often above those in the ambient air. Moreover, the energy utilized by the photosynthetizing cell is that of light. This energy, mainly at the ultraviolet and visible regions, can be harmful by itself, even without the risks involved with activation of molecular oxygen.

It is the combination of these two life-giving effectors, light and oxygen, which sometimes turn to be damaging and even deadly to oxygenic organisms. Various terms have been coined over the years to describe this combined effect: photooxidation, photoinhibition, photoinactivation, photolability, and more.

On the other hand, being more susceptible to photooxidative conditions, today's oxygen producing cells are those that by evolutionary necessity have developed the apparatus to cope, at least up to a certain point, with the dangers involved in exposure to light and oxygen.

These two aspects of oxygen toxicity — the damage and the protection against it — are reviewed and discussed in some detail by several of the references in the "Suggested Reading" list at the end of this chapter. Most of them, however, are either general in character or discuss the matter as it manifests itself in the higher plant's chloroplast. In this chapter we would like to deal mainly with the "oxygen problem" as it is faced by a single group of photosynthetic organisms — the cyanobacteria (blue-green algae).

Members of this group share with eukaryotic algae and higher plants the complete, O_2-evolving photosynthetic apparatus. Nevertheless, they are prokaryotic in all respects. They are also among the most ancient organisms on earth (fossils of cyanobacteria-like organisms have been found in rocks dated to be 3×10^9 years old); in fact, the ancient forefather of the modern chloroplast may very well have been a cyanobacterium. More importantly, the cyanobacteria date back to times in which earth was anaerobic; this is still evidenced by the fact that many cyanobacteria are also capable of carrying out a bacteria-type, anaerobic, anoxygenic photosynthesis. Cyanobacteria had therefore to evolve and develop the means to cope with the transfer from an O_2-free world to the mostly O_2-dependent biosphere of today. This evolutionary record is hidden in the intricacies of cyanobacterial structure and metabolism, and is being continuously unraveled. In this chapter, however, we mainly wish to present the picture as it is today — how cyanobacteria cope with the harmful effects of molecular oxygen and its active species.

MEASUREMENT OF DISSOLVED OXYGEN

Total Oxygen

The factors involved in O_2 toxicity and in protection against it are many. One of the most obvious parameters is the concentration of dissolved oxygen — the actual amount of O_2 "seen" by the affected cell.

Numerous methods exist for the quantitative assay of dissolved O_2, the most common of which is the polarographic one, or the oxygen electrode. Recently, a completely novel approach has been emerging, utilizing ESR spectroscopy for the same purpose. Several unique features of this methodology make it suitable for the study of the effect of O_2 on the living cell. Not only can the concentration of dissolved O_2 be measured in the same experimental systems in which oxidative damage is being monitored by other means, but also, more importantly, for the first time we are given the tool to measure *intracellular* O_2 levels.

The approach makes use of nitroxides, stable radicals which have been used as ESR spin probes for many other purposes. One of the characteristic parameters of a spin probe signal is its line width. As can be seen in Figure 1, a typical nitroxide spectrum consists of three lines. The width of a line is the horizontal distance, in units of magnetic field (Gauss or Tesla), between its peak and trough. This parameter is sensitive to various environmental effectors, among them the medium's viscosity and the presence of other paramagnetic species, which cause line broadening due to collision-dependent spin exchange. The degree of broadening is directly proportional to the concentration of the paramagnetic agent. Since molecular oxygen is paramagnetic, measurements of line broadening or narrowing upon changes in O_2 concentration should lead to quantitation of the latter.

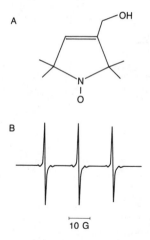

Figure 1. Chemical structure (A) and ESR spectrum (B) of the nitroxide spin probe 2,2,5,5-tetramethyl-3-methonolpyrroline-1-oxyl (PCAOL).

Two main approaches have been introduced in order to carry this out. In the first, it is not the width of the line itself, as earlier described, which is followed. The parameter monitored is the height of the superhyperfine lines. These are sort of "distortions" in the main line shape of the spectrum, caused by interactions between protons and the unpaired electron. Upon the introduction of oxygen and the subsequent signal broadening, the superhyperfine lines "flatten out". This latter phenomenon and several variations, have been utilized to very precise quantitation of the amount of oxygen present (Francisz et al., 1985; Lai et al., 1982). O_2 concentrations as low as 0.1 μM, in samples not larger than 1 μl, have been readily measured (Morse & Swartz, 1985)). The superhyperfine lines tend, however, to "flatten out" at relatively low O_2 concentrations. This does not matter if these still fall within the desired physiological range of the system studied. For higher O_2 levels, for instance above air-saturation (approximately 0.25 mM), this approach is of limited value.

The second approach is based upon measurements of the width of the line itself, and not of its superhyperfine additions. Sensitivity is somewhat decreased, but the range is multiplied many-fold. It is thus more applicable to photosynthetic systems in which O_2 can reach relatively high levels. Figure 2 presents an expanded midfield line spectrum of the nitroxide spin probe, PCAOL, in the presence of a dense suspension of the cyanobacterium *Agmenellum quadruplicatum*. From a width of 1.4 Gauss in the dark, under O_2-free conditions, the signal broadens to 1.9 Gauss after 1 minute of illumination.

If such measurements are conducted sequentially at closer intervals, before, during, and after illumination, plots such as the one presented in Figure 3 can be generated. Upon cessation of illumination, the lines narrow again to their initial 1.4 Gauss width, and the procedure can be repeated several times.

In Figure 3, the changes in line width are quantitated not only by magnetic field units (left hand side), but also by actual O_2 concentrations (right hand side). The latter

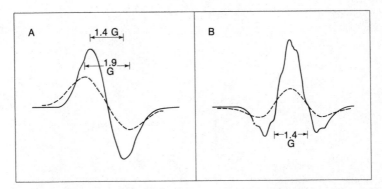

Figure 2. The midfield line of the PCAOL spectrum on an expanded scale, in the presence of a dense cyanobacterial suspension (1 mg chl · ml^{-1}). A) the original signal (1st derivative), B) 2nd harmonic. Solid line (—) before illumination. Dashed line (---) after 1 min in the light.

can be easily determined by measuring the line widths of probe solutions saturated by N_2 gas containing various concentrations of O_2. From the known solubility of O_2 at the appropriate temperature, pressure, and salinity, the concentrations of dissolved O_2 for each of those gas mixtures can be derived, and can be correlated with the line widths measured. For the experiment described in Figure 3, for instance, a calibration factor of 0.254 G/mM O_2 was calculated.

Rather than repeatedly scanning the midfield line, measuring its width each time, and then plotting the results, a more convenient option exists. The ESR spectrum, as

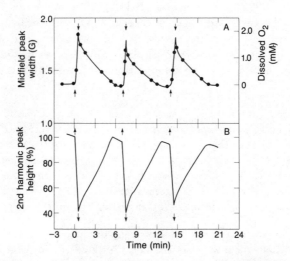

Figure 3. Fluctuations in O_2 concentration in a cyanobacterial preparation, (1 mg chl · ml^{-1}) during a series of 30 sec illuminations, as monitored by PCAOL (0.5 mM). A) direct line width measurement. B) continuous traces of the 2nd harmonic. Upward arrows — light on. Downward arrows — light off.

we usually see it, is actually a first derivative. ESR spectrometers also allow the plotting of the 2nd derivative, or the 2nd harmonic. Figure 2 also shows traces of the 2nd harmonic of the central peak before and after a 1 min illumination. Upon broadening of the 1st derivative line, the slope between its peak and trough decreases. The height of the 2nd harmonic faithfully follows changes in the slope, and is substantially reduced. By locking the magnetic field into the peak of the 2nd harmonic signal, and following its height during a series of 30 sec illuminations, an apparent "mirror image" of the broadening curve in Figure 3a is obtained (Fig. 3b).

It should be noted that whereas the use of the 2nd harmonic peak as a monitor of O_2 concentrations is convenient for kinetic measurements and rapid O_2 transients, its quantitation is not straight forward. A derivative approach is non-linear, and careful calibration should be carried out. In addition, at approximately 2 mM O_2, peak height is so small that most of the sensitivity is lost.

Intracellular Oxygen

A unique feature of ESR spectroscopy of spin probes is the possibility to mask all signals emanating from radicals outside the cell, and therefore to "see" only the intracellular signals. This can be achieved by the addition of a chemical spin probe broadening agent, usually a transition metal complex. Unlike the completely permeable spin probe, this quenching agent should be impermeable. Upon its addition, therefore, the only ESR-detectable signal would be that generated by the probe molecules inside the cells. Thus, the width of the observed signal is in direct correlation with the internal O_2 concentration. An example of such a measurement is presented in Figure 4, which depicts the increase and decline in O_2 concentrations inside *A. quadruplicatum* during and following a 45 sec illumination. The calibration factor used here was different than before; the increased internal viscosity affects the degree of broadening, whereas the internal solutes and turgor pressure may affect O_2 solubility. For the experiment described in Figure 4, a broadening calibration factor of 0.189 G/mM O_2 was calculated.

This approach can also be applied to the method that uses the hyperfine structure of the spin probe spectrum for the determination of internal O_2 concentrations (Sarna et al., 1980), with high precision but a limited range, as discussed before.

With a proper calibration procedure, the above methodology can be applied to any spin probe; line width measurements can be carried out during any other ESR assay involving spin probes. It can also be utilized directly for the measurement of the oxidative conditions in many of the systems described below.

PROTECTION AGAINST HARMFUL EFFECTS OF OXYGEN AND LIGHT

In the protective strategy of cyanobacteria against the cumulative effects of O_2 and light, several levels can be discerned.

The first of these may be termed "behavioral", though it is, of course, physiological in nature. In a mode generally analogous to leaf motions in the light, or to chloroplast

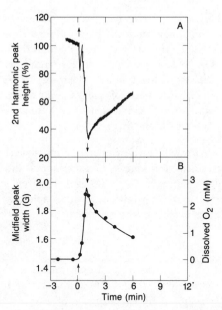

Figure 4. Fluctuations in cyanobacterial intracellular O_2 concentrations, during and after illumination (45 sec), as monitored by PCAOL (2 mM). A) 2nd harmonic traces. B) direct line width measurements. Upward arrow — light on. Downward arrow — light off. Chlorophyll concentration — 1 mg chl · ml^{-1}.

movement within the plant cell, some cyanobacteria can adjust their vertical position in their aquatic habitats. In that manner, optimal light and/or O_2 conditions can be reached, and the "danger-zones" avoided. The cyanobacterial movements are caused by changes in buoyancy, which in turn is regulated by the osmotic effect of metabolites on gas vesicles within the cell. Thus, for a gas vacuolated organism, high photosynthetic activity would result in an increased internal concentration of metabolites, thereby increasing turgor pressure; the gas vesicles would be compressed and at least partially collapse; buoyancy will be lost, and the cells will sink. Once the cells are in relative darkness, the opposite will occur, finally leading to a position most suitable for the cells. In fact, in many water bodies containing cyanobacteria, the latter may be found in a relatively narrow layer, migrating down and up as light intensity increases and then decreases during the day. This movement, in addition, may also be regulated in some cases by the availability of nitrogen sources, mainly ammonia.

The above strategy is restricted to gas vacuolated cyanobacteria. Furthermore, an "escape" mechanism, efficiently regulated as it may be, is only a partial answer for dealing with excess light and O_2. Other ways of dealing with these problems involve the structural or "biophysical" characteristics of these organisms. The presence of carotenoids in photosynthetic membranes, for instance, is believed to play a protective role, dissipating the excitation energy as heat, and possibly also serving as a sink for oxygen radicals. Other biophysical means may be nondestructive energy dissipation as fluorescence, and the effect known as "spillover" of electrons from photosystem II to photosystem I. Both of these may limit the amount of light reaching the photosynthetic reaction centers, but have not been shown to actually prevent photo-inactivation.

All of these strategies may indeed serve to limit the amounts of reactive oxygen species; it is the metabolic pathways, however, that help the cell "fight for its life". Indeed, superoxide, hydrogen peroxide, hydroxyl radicals, singlet oxygen, and organic peroxides have all been shown or suggested to occur in photosynthetic systems. In cyanobacteria, most of the accumulated data are those concerned with dealing either with the superoxide radical or with H_2O_2.

Superoxide, which can be generated by photosystem I activity in cyanobacteria as in chloroplasts, may dismutate into O_2 and H_2O_2:

$$O_2^- \cdot + O_2^- \cdot + 2H^+ \rightarrow H_2O_2 + O_2 \tag{1}$$

This reaction may occur spontaneously, or catalyzed by the enzyme superoxide dismutase (SOD). The importance of this reaction may be evidenced by the observation that it seems to be obligatory for all aerobic or aerotolerant organisms. In fact, the presence of various forms of the enzyme (differing mainly in their metal content) in different types of organisms have served as a basis for several evolutionary schemes.

In cyanobacteria, it is mainly the iron-containing enzyme that has been shown to correlate with the ability of various strains to adapt to O_2-rich conditions. Levels of SOD found in laboratory grown strains or natural populations of cyanobacteria correlated with their degree of susceptibility to photooxidations. SOD levels were higher in resistant strains; when, due to various treatments, these levels dropped to about 10% of their normal values, photooxidative death almost invariably occurred.

H_2O_2 can be produced in cyanobacteria by the activity of SOD, as described above, as well as independently by photorespiration, and may be removed by several metabolic pathways. Possibly the better known enzyme in that group is catalase, which uses H_2O_2 as a substrate to produce H_2O and O_2 :

$$2H_2O_2 \xrightarrow{\text{catalase}} 2H_2O + O_2 \tag{2}$$

The activity of catalase in different cyanobacterial strains varies, but is often low. There are also indications that the enzyme might be light-inactivated. A complementary system has therefore to exist in order to encounter the total potential of H_2O_2 production in cyanobacterial cells. This system has recently been unraveled by Tel-Or et al. (1986), using two cyanobacterial strains, *Nostoc muscorum* and *Synechococcus* 6311. Their findings will be discussed below.

In chloroplasts, some of the H_2O_2 is scavenged by ascorbate peroxidase. As its name implies, this enzyme peroxidizes ascorbate to dehydroascorbate (DHA) using H_2O_2. However, whereas in chloroplasts internal ascorbate concentrations were found to be high (up to 25 mM), the levels of this reductant in the two strains mentioned above did not exceed 0.1 mM. These very low concentrations, though at first glance seemingly insufficient, may suffice for efficient H_2O_2 removal, providing that two conditions are met:

high ascorbate peroxidase activity and a fast regeneration of the reduced ascorbate. These two requirements have been found to be fulfilled in these two cyanobacteria. The enzymatic activity is indeed high, and ascorbate is continuously regenerated from DHA by DHA-reductase.

This cascade of events, however, does not stop at this point, for DHA-reductase requires an electron donor for its reducing activity. These electrons are apparently provided by glutathione (GSH), which is oxidized in the process to glutathione disulfide (GSSG). The latter, in turn, can be returned to its reduced form by the NADPH-dependent activity of GSSG-reductase. The complete scheme, as emerging from these studies, is presented in Figure 5.

It should be noted that ascorbate peroxidase is not the only peroxidase involved in H_2O_2 removal. In mammalian systems, it is GSH-peroxidase which plays the major role in cytoplasmic H_2O_2 scavenging. In the studies summarized in Figure 5, however, activity of this enzyme was not detected even though GSH levels were appreciable.

Another point to be remembered is that like in other biological systems, SOD, catalase, and the peroxidases are not the only metabolic means for protection against the deleterious effects of O_2 activation. Small molecules such as glutathione, ascorbate, hydroquinones, or alpha-tocopherol may react directly or enzymatically with $O_2^-\cdot$, H_2O_2 or OH$\cdot$, while carotenes, unsaturated fatty acids, and alpha-tocopherol have been shown to react directly with singlet oxygen. Except for the protective presence of carotene, which has been discussed earlier, none of the other reactions just mentioned have been studied in cyanobacteria.

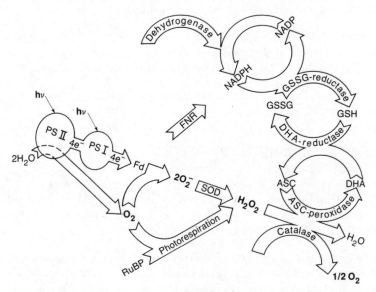

Figure 5. H_2O_2 production and removal in cyanobacteria. ASC — ascorbate; FD — ferredoxin; DHA — dehydroascorbate; FNR — ferredoxin-NADP reductase; GSH — glutathione; GSSG — glutathione disulfide; PS I, II — photosystem I, II; RUBP — ribulose bisphosphate. From reference 5.

N₂ FIXATION AND THE HETEROCYST

Many cyanobacteria, in addition to their "higher plant" photosynthesis, can also fix atmospheric nitrogen — an exclusively prokaryotic process. This presents a most intriguing problem, since the nitrogenase enzyme complex "responsible" for N_2 fixation is drastically O_2-sensitive, and is irreversibly inactivated within a few seconds of exposure to air. The nitrogenase should therefore be protected from the presence of even trace amounts of molecular O_2.

Asimbiotic, N_2-fixing cyanobacteria can be divided into three groups in regard to the aerobic activity of the nitrogenase. In members of the first group, the nitrogenase is active only under anaerobic or microaerobic conditions. In other words, they do fix N_2, but only in the absence of O_2. The two other groups, however, can actively fix N_2 under aerobic conditions. The second group consists of heterocystous cyanobacteria; when the availability of fixed nitrogen declines, these filamentous organisms differentiate specialized cells — the heterocysts — in which nitrogen fixation takes place. O_2 levels in the heterocysts are apparently kept to a minimum by three independent mechanisms: 1) The heterocysts lack photosystem II activity, and therefore do not evolve O_2; nitrogen fixation and O_2 production are thus spatially separated; 2) there are indications that the thick, multi-layered heterocyst cell wall slows down the inward diffusion of molecular oxygen; 3) respiration in the heterocysts is increased to a level sufficient for dealing with the O_2 that does get in. Interestingly, SOD levels in the heterocysts appear to be much lower than in the vegetative cells, at least in *Anabaena cylindrica*. Apparently in a cell where the presence of molecular O_2 itself may mean instant loss of activity, protection against superoxide radicals is somewhat superfluous. Catalase activity, however, was found to be very similar in both cell types.

To the third cyanobacterial group belong the nonheterocystous, aerobic nitrogen-fixers. Both O_2 evolution and N_2 fixation appear to take place in the same cells, thereby posing, probably, the greatest challenge for protecting the nitrogenase. As a matter of fact, the manner in which this is accomplished has not been completely resolved to date, although several possibilities have been suggested. These include temporal separation of the two activities, enhanced respiration and the hydrogenase catalyzed O_2-consuming oxy-hydrogen reaction, and high turnover of nitrogenase peptides.

ACKNOWLEDGMENTS

L. Packer has collaborated with Elisha Tel-Or and Margaret Huflejt in the cyanobacterial studies of hydroperoxide removal, and both authors with Rolf Mehlhorn and Pedro Candau in developing the ESR spin label oximetry method for measuring O_2 concentrations in cyanobacteria. Original research described in this chapter was supported by the Office of Biological Energy Research, Department of Energy (Grant No. W-7405-ENG-48), the National Foundation for Cancer Research, and the National Aeronautics and Space Administration (Controlled Ecological Life Support Systems Program, Grant No. NCA2-ORO-401). S. Belkin is a recipient of the Dr. H. Weizmann Post-Doctoral Fellowship for Scientific Research.

REFERENCES

Francisz, W., Lai, C.-S., and Hyde, J.S., 1985, *Proc. Natl. Acad. Sci.*, 82:411-415.

Lai, C.-S., Hopwood, L.E., Hyde, J.S., and Lukiewica, S., 1982, *Proc. Natl. Acad. Sci.*, 79:1166-1170.

Morse, P.D. and Swartz, H.M., 1985, *Magn. Reson. Med.*, 2:114-127.

Sarna, T., Duleba A., Korytaski, W., and Swartz, H., 1980, *Arch. Biochem. Biophys.*, 200:140-148.

Tel-Or, E., Huflejt, M., and Packer, L., 1986, Hydroperoxide metabolism in cyanobacteria., *Arch. Biochem. Biophys.*, *245*, in press.

SUGGESTED READING

Berliner, L.J., ed., "Spin Labelling, Theory and Applications," Vol. I, 1976, Vol. II, 1979. Academic Press, New York, San Francisco, London.

Elstner, E.F., 1982, Oxygen activation and oxygen toxicity, *Ann. Rev. Plant Physiol.* 33:73-96.

Halliwell, B., 1984, "Chloroplast Metabolism: The Structure and Function of Chloroplasts in Green Leaf Cells," Clarendon Press, Oxford.

Halliwell, B., and Gutteridge, J., 1985, "Free Radicals in Biology and Medicine," Clarendon Press, Oxford.

Powles, S.B., 1984, Photoinhibition of photosynthesis induced by visible light, *Ann. Rev. Plant Physiol.*, 35:15-44.

Michelson, A.M., McCord, J.M., and Fridovich, I., eds., 1977, "Superoxide and Superoxide Dismutases," Academic Press, London.

Van Liere, L. and Walsby, A.E., 1982, Interactions of cyanobac- teria with light, in "The Biology of Cyanobacteria," N.G. Carr and B.A. Whitton, eds., pp. 9-45, University of California Press, Berkeley, Los Angeles.

DETECTION AND CHARACTERIZATION
OF SINGLET OXYGEN

Christopher S. Foote

Department of Chemistry and Biochemistry
University of California
Los Angeles, California 90024

INTRODUCTION

Light and oxygen are toxic. Environmental chemicals or natural cell constituents that absorb light (such as porphyrins or flavins) can "sensitize" organisms to damage. Examples in man include photosensitive porphyrias, drug photosensitivity, and photoallergy. Aging of sun-exposed skin, cataract induction, and photocarcinogenesis may be caused by related chemical mechanisms. Damage to organisms caused by light and oxygen in the presence of dyes or pigments is called "photodynamic action"; damage to biological target molecules includes enzyme deactivation (through destruction of specific amino acids, particularly methionine, histidine, and tryptophan), nucleic acid oxidation (primarily guanine), and membrane damage (unsaturated fatty acids and cholesterol are targets). There are many naturally-occurring photodynamic sensitizers that can cause harm to humans, mammals, or plants: for example, porphyrins from blood or chlorophyll, the plant toxin hypericin, the fungal pigment cercosporin, and several polyacetylene derivatives.

MECHANISMS OF PHOTOOXYGENATION

Photosensitized oxidations are initiated by absorption of light by the sensitizer, which can be a dye or pigment, a ketone or quinone, an aromatic molecule, or many other types of compounds. The "sensitizer" (Sens) is converted to an electronically excited state (Sens*) by absorption of a photon. The initial product is a short-lived singlet (^{1}Sens); in many cases, this undergoes intersystem crossing to the longer-lived triplet (^{3}Sens). Because of its short life, only materials at relatively high concentration can interact with the singlet; however, much lower concentrations can still react with the longer-lived triplet state.

$$\text{Sens} \xrightarrow[h\nu]{} {}^{1}\text{Sens} \longrightarrow {}^{3}\text{Sens}$$

There are two mechanisms of sensitized photooxidation, called "Type I" and "Type II" by Gollnick and Schenck (see scheme below). In the Type I process, substrate or solvent reacts with the sensitizer excited state to give radicals by hydrogen atom or electron transfer. Reaction of these radicals with oxygen gives oxygenated products. In the Type II process, the excited sensitizer reacts with oxygen to form singlet molecular oxygen ($^{1}O_2$), which reacts with substrates to give oxygenated products. Factors which control the competition between Type I and Type II mechanisms include oxygen concentration, the reactivities of the substrate and of the sensitizer excited state, and the substrate concentration. A complete discussion of these factors is beyond the scope of this paper.

Sens

Type I $\downarrow h\nu$ Type II

Radicals $\longleftarrow$ Sens* $\longrightarrow$ $^{1}O_2$

Substrate or Solvent O_2

$\downarrow O_2$

Oxygenated Products   Oxygenated Products

THE TYPE I PROCESS

Electron transfer both to and from molecules takes place more readily in the excited state than in the ground state. This follows from the fact that in the excited state, an electron has been promoted from an orbital which is strongly binding in the ground state to one which is less strongly binding; this process results in a strongly reducing electron and a strongly oxidizing hole.

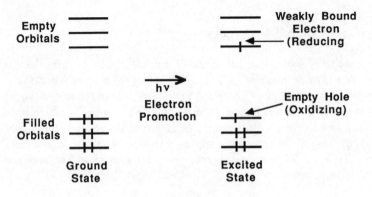

An example of a Type I photosensitized oxidation is the photooxidation of aromatic olefins (Donor) such as stilbene, sensitized by electron-poor aromatics such as dicyanoanthracene (DCA) in their excited singlet state. This reaction has been shown to proceed by the electron-transfer mechanism shown below to produce initial radical ions; the radical anion of DCA is reoxidized by oxygen, producing superoxide ion, which reacts with the substrate radical cation, producing the product, in this case, benzaldehyde.

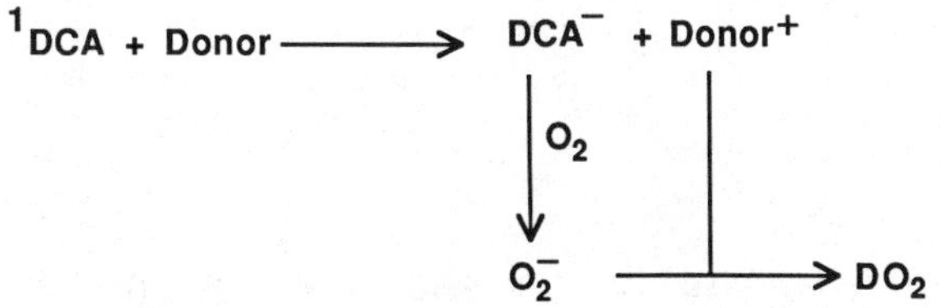

$$^1DCA + Donor \longrightarrow DCA^- + Donor^+$$

$$\downarrow O_2 \qquad \qquad \qquad |$$

$$O_2^- \longrightarrow DO_2$$

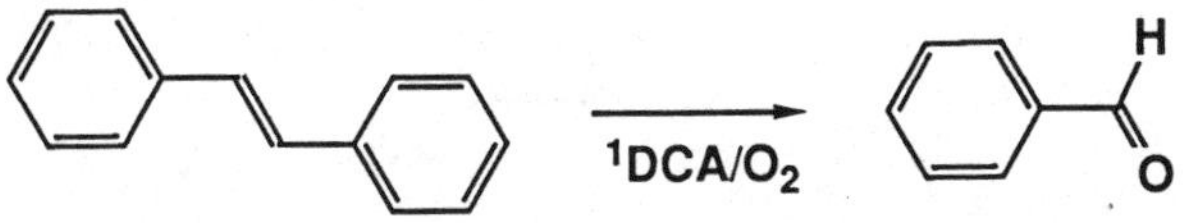

The Type I reaction can also lead to hydrogen abstraction, which can lead to the formation of peroxides, radical chain autooxidation, and other consequences. This reaction is particularly common with ketones, but also occurs with many dyes, although usually less efficiently. Good hydrogen donors promote it.

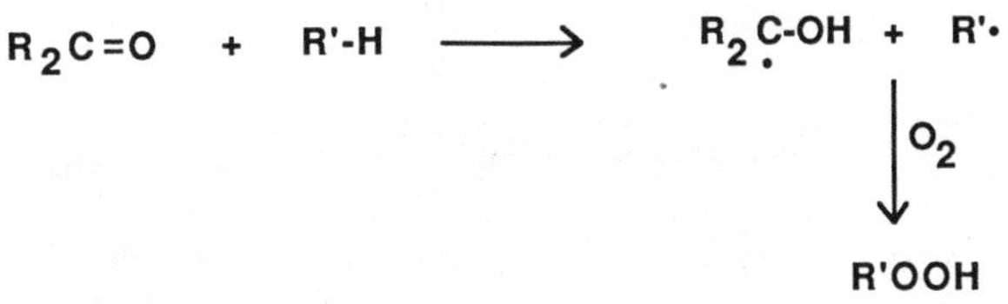

$$R_2C=O + R'-H \longrightarrow R_2\overset{.}{C}-OH + R'\cdot$$

$$\downarrow O_2$$

$$R'OOH$$

THE TYPE II PROCESS

The Type II reaction produces oxidized compounds from the reaction of singlet oxygen with the substrate. Two classes of reaction of singlet oxygen are best known: addition to dienes to produce endoperoxides, and additions to olefins, giving allylic hydroperoxides.

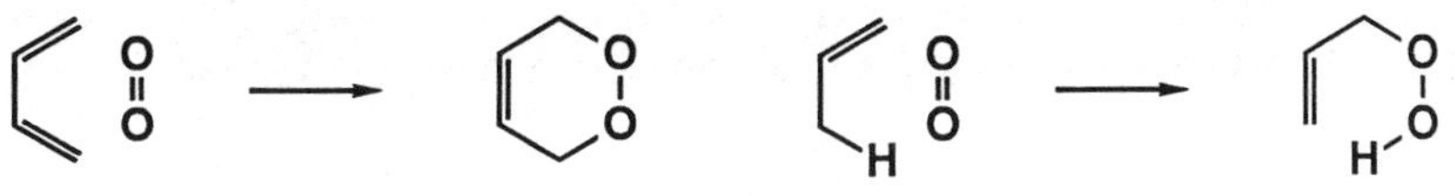

109

In addition, electron-rich olefins react with singlet oxygen to produce dioxetanes, unstable four-membered ring peroxides which often break down with the emission of light. Sulfides are oxidized to sulfoxides via an unstable intermediate (R_2SOO); phenols can also be oxidized, producing unstable hydroperoxides.

$$^1O_2 + R_2S \longrightarrow R_2SOO \longrightarrow 2\,R_2SO$$

Many compounds quench (deactivate without reaction) singlet oxygen efficiently. For example, β-carotene inhibits photooxidation of 2-methyl-2-pentene efficiently at 10^{-4} M by an energy transfer mechanism, without being appreciably oxidized itself. Other types of compounds such as DABCO (1,4-diazabicyclooctane) and azide ion also quench singlet oxygen, but by a charge transfer process. Phenols and sulfides also quench singlet oxygen in competition with their reaction. For example, α-tocopherol quenches singlet oxygen at a high rate in all solvents, but reacts rapidly only in protic solvents.

$$^1O_2 + Car \longrightarrow {}^3Car + {}^3O_2$$

$$^1O_2 + DABCO \longrightarrow O_2^{\cdot -}\cdots DABCO^+ \longrightarrow DABCO + {}^3O_2$$

Singlet oxygen was first detected by its luminescence. Emission from singlet oxygen in solution is extremely inefficient. There are two types: luminescence from a single molecule at 1.27μm, and "dimol" luminescence at 634 and 704nm. The infrared luminescence is very weak because the lifetime of singlet oxygen in solution is short. The dimol efficiency also depends on the concentration of singlet oxygen, since a bimolecular collision between two short-lived species is required.

$$^1O_2 \longrightarrow h\nu\ (1.27\mu) \qquad 2\,(^1O_2) \longrightarrow h\nu\ (634,704nm)$$

A short pulse of laser light can be used to excite singlet oxygen sensitizers; the 1.27μm luminescence can be detected by a germanium photodiode with a low-noise amplifier and a digitizer with signal averaging. The amount of singlet oxygen produced and its lifetime can be measured very easily this way. This technique provides the most definitive and quantitative method of characterizing singlet oxygen produced in photochemical systems. Furthermore, by measuring the change in lifetime of $^1O_2$ when a reagent is added, the rate of its reaction with $^1O_2$ can be simply and rapidly determined.

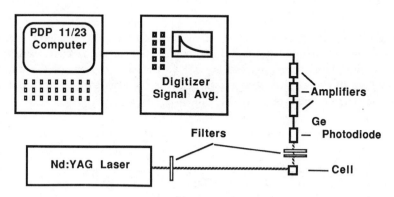

Figure 1. Singlet O_2 Emission.

Establishing the mechanism of photosensitized oxidations in complex systems is difficult. Reaction kinetics, comparison of products with those of known singlet oxygen reactions, the quantitative effect of quenchers, direct detection of singlet oxygen and various excited species and measurement of their reactions by time-resolved techniques, or a combination of these techniques must be used. However, many other strong oxidants may also be present, and distinguishing their fingerprints from those of singlet oxygen may be difficult. Similar problems obtain when the singlet oxygen source is nonphotochemical.

NONPHOTOCHEMICAL REACTIONS

Singlet oxygen can also be formed without light from hypochlorite and hydrogen peroxide, by decomposition of peroxides, and in other reactions. Many reports have claimed the detection of singlet oxygen in photochemical and thermal systems, especially biological. However, other oxidizing species can be confused with singlet oxygen if tests are not specific; specific identification of the reactive species in these systems may be even more difficult than in photochemical systems, since time-resolved techniques can usually not be used. Possible reactive species include $^1O_2$, superoxide ion $(O_2^-\cdot)$ and stronger oxidants derived from it, (often by interaction with hydrogen peroxide, the product of its dismutation, catalyzed by iron species), hydroxy radical (OH·), alkoxy radicals (RO·), peroxy radicals (ROO·), and other oxidants such as hypochlorous acid.

What is needed are methods of assessing the relative importance of various processes. It does little good to show that a reactive intermediate is present without being able to estimate what fraction of the overall oxidation it causes. The next section surveys a few techniques which are available for characterizing singlet oxygen, emphasizing those which can be quantified.

TECHNIQUES FOR CHARACTERIZING SINGLET OXYGEN

Chemical Traps

A large number of compounds that react with singlet oxygen have been added to reacting systems as traps, and the formation of the supposedly characteristic products used as an indication of the intermediacy of $^1O_2$. For example, dimethylfuran reacts with singlet oxygen to give the diketone shown below as the ultimate product. Unfortunately, so do a very large number of other oxidants.

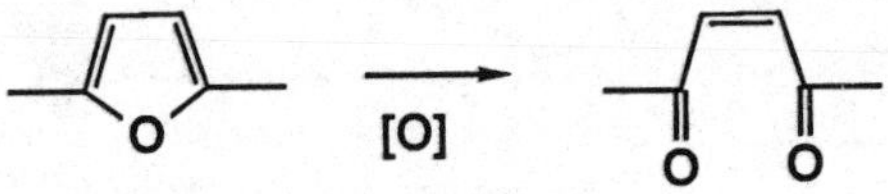

One trap which does appear to be diagnostic for singlet oxygen is cholesterol, which reacts with singlet oxygen to give the 5-β-hydroperoxide; reactions with radical and other oxidants give complex mixtures, but the 5-β-product is not among them. This system is somewhat limited because of the low reactivity of cholesterol with singlet oxygen. Although cholesterol is not soluble in water, it can be bound to microspheres, allowing its use in aqueous systems.

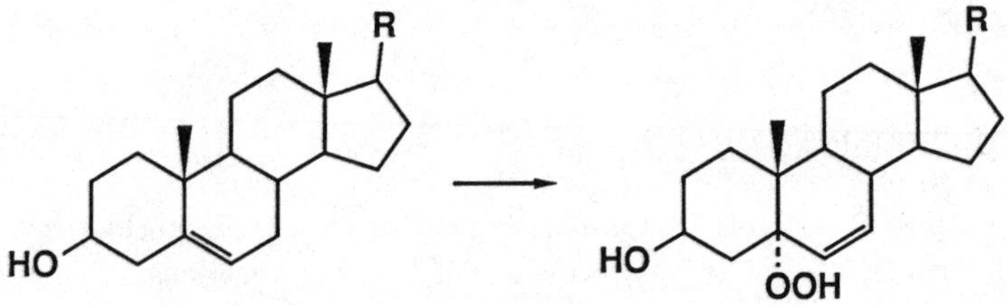

A second trapping system which has been used in several cases, and which also appears to be specific, is a suitably substituted anthracene. Anthracene derivatives are considerably more reactive than cholesterol. These compounds can be made soluble in any medium by suitable choice of substituents. One drawback to this system is that anthracenes are also photosensitizers, so that when small amounts of product are formed, adventitious photooxidation must be carefully ruled out.

A third trapping system makes use of the fact that polyunsaturated fatty acids react with singlet oxygen to give a mixture of conjugated and unconjugated isomers of the product hydroperoxides, whereas only the conjugated isomers are formed on radical attack. This system, like the cholesterol trap, is somewhat difficult to use, since the isomers must be separated by HPLC.

Inhibitors. As mentioned above, many compounds such as carotene, DABCO, and azide, are effective quenchers for singlet oxygen. These compounds, and others which react with singlet oxygen, are frequently used to inhibit reactions in which singlet oxygen is thought to be a reactive intermediate. Because of their lack of specificity, care must be taken in interpretation of the results, since all are compounds of low redox potential which will certainly react with any other strong oxidants present. One way of using inhibitors that partly avoids this problem is to use a quantitative treatment, calculating the amount of singlet oxygen expected to be inhibited from known rate constants and comparing it with that observed. This technique cannot be used in inhomogeneous solutions, where the local concentration of the inhibitor cannot be calculated.

D_2O Effect. Singlet oxygen has a longer lifetime in D_2O than in H_2O (Kearns, 1979; Lindig and Rodgers, 1981). Thus many reactions of singlet oxygen proceed more efficiently in D_2O than in H_2O. However, there are two important limitations to this technique. First, singlet oxygen reactions in the two solvents will differ in efficiency only if solvent decay of singlet oxygen is rate limiting; if substrate or quencher is removing all the $^1O_2$, there will be no effect on the lifetime. Secondly, it has been shown that $O_2^-\cdot$ also has a longer lifetime in D_2O than in H_2O; the effect of solvent deuteration on other possible reactive species has not been shown. Thus, this effect, by itself, cannot be used to distinguish between reactions of $^1O_2$ and $O_2^-\cdot$, and is not a strong argument for the intermediacy of $^1O_2$ in the absence of strong corroborating evidence.

Luminescence. Dimol (visible) luminescence may be specific for singlet oxygen, but can not be easily used to determine the amount of singlet oxygen present since it depends on a second-order reaction between two singlet oxygen molecules. It is essential that the wavelength of the emission be carefully determined; many reports of singlet oxygen production by this technique have been misleading, because the source of light emission was subsequently found to be something other than singlet oxygen when its wavelength was determined. The high sensitivity of many photomultipliers also means that tiny amounts of light can be measured, which may have little relationship to chemical processes going on. Direct observation of the infrared luminescence of singlet oxygen can lend confidence to its identification if the wavelength is carefully established. This technique has been applied to nonphotochemical systems in only a few cases so far.

TECHNIQUES FOR CHARACTERIZING OTHER REACTIVE OXYGEN SPECIES

Techniques which are similar to those used for singlet oxygen (characteristic product formation, inhibition by "specific" inhibitors, and a few more specialized techniques) can be used for other reactive oxygen species mentioned above. However, most of these techniques have not been worked out to the point where they can be used to quantify the presence of these species. Because the chemistry of these species is discussed in detail in other chapters in this book, further discussion will be omitted.

"CLEAN" SOURCES OF SINGLET OXYGEN

One often wishes to know how to generate singlet oxygen under carefully defined conditions and free of any other reactive species. Photochemical systems (using nonreactive sensitizers, at high O_2 pressure, and with low concentrations of substrates that are unreactive in the Type I reaction) can often be used. Another technique is to use a reverse Diels-Alder reaction, using a naphthalene or anthracene endoperoxide; this technique can be used under very mild conditions, and no side reactions have yet been reported. Most other known chemical sources of singlet oxygen (e.g., hypochlorite/H_2O_2, phosphite ozonides) involve the use of very strong oxidants which can interfere with the specific reactions of singlet oxygen.

EXAMPLE: POLYMORPHONUCLEAR LEUKOCYTES

The polymorphonuclear leukocyte (PMN) system is a particularly interesting example of the use of various techniques for trapping reactive intermediates. These cells have an essential oxygen-dependent antimicrobial system which is often said to involve the production of "oxygen radicals" as toxic agents. Recently, considerable interest has been excited by the demonstration that PMN's produce large amounts of strong chlorinating species. Oxygen uptake is followed by reduction to superoxide, then to hydrogen peroxide; the hydrogen peroxide is used by a myeloperoxidase (MPO) to oxidize chloride ion to the Cl^+ level.

$$O_2 \longrightarrow O_2^- \longrightarrow H_2O_2 \xrightarrow[MPO]{} HOCl$$

Our interest in these cells began with the suggestion that singlet oxygen from the reaction of hypochlorite ion and H_2O_2 was the antimicrobial agent. We used a trap designed to give a singlet oxygen fingerprint to test this suggestion. Radiolabeled cholesterol, which, as mentioned above, gives a hydroperoxide specific for singlet oxygen, was adsorbed on polystyrene microbeads. These beadlets are phagocytized by the PMN's. After incubation, the cells were broken up, the steroid products were extracted, and the mixture chromatographed along with various known products, which were isolated and their radioactivity assayed. No evidence for the production of singlet oxygen was found. Although this experiment was difficult to make quantitative, the absence of the singlet oxygen product suggested that singlet oxygen could not be the major antimicrobial species.

Instead of the hydroperoxide, small amounts of epoxide were found, which proved to be identical to product produced from cholesterol under the action of HOCl or $MPO/H_2O_2/Cl^-$. Since our original system had not been designed to test for chlorinating species, a reactive trap for Cl^+ was needed. Trimethoxybenzene (TMB) was found to react rapidly with either HOCl or $MPO/H_2O_2/Cl^-$ to give the monochloro derivative. TMB was then used in the PMN system as an assay for chlorinating species; it is chlorinated by PMN's in yields of 25-50%, based on oxygen uptake. This experiment shows that large amounts of a chlorinating species are formed by activated PMN's, as reported by several other groups. It is not yet certain what the halogenating species is; HOCl is possible, but an important role of chloramines has also been suggested.

SUGGESTIONS FOR FURTHER READING

Photochemistry

Turro, N.J., 1978, "Modern Molecular Photochemistry," Benjamin/Cummings, Menlo Park, Ca.

Photodynamic Action

Krinsky, N.I., 1983, Biological roles for singlet oxygen, in "Singlet Oxygen," H.H. Wasserman and R.W. Murray, Eds., Academic Press, New York, 597.

Spikes, J.D., 1982, Photodynamic reactions in photomedicine, in "The Science of Photomedicine," J.D. Regan and J. A. Parrish, eds., Plenum Publishing Corp., New York, 113.

Singlet Oxygen

H.H. Wasserman and R.W. Murray, eds., "Singlet Oxygen," Academic Press, New York.

Characterization of Reactive Intermediates

Foote, C.S., 1979, Detection of singlet oxygen in complex systems: a critique, in "Biochemical and Clinical Aspects of Oxygen," W.S. Caughey, ed., Academic Press, New York, 603.

Foote, C.S., 1985, Dicyanoanthracene sensitized photooxygenation of olefins: Electron Transfer and Singlet Oxygen Mechanisms, *Tetrahedron*, 41:2221.

Packer, L., Ed., 1984, Oxygen radicals in biological systems, in *Methods in Enzymology*, 105: .

Singh, A., 1982, Chemical and biochemical aspects of superoxide radicals and related species of activated oxygen, *Can. J. Physiol. Pharm.*, 60:1330.

Leukocytes

Babior, B. M., 1984, Oxidants from phagocytes: agents of defense and destruction, *Blood*, 64: 959.

BIOLOGICAL CHEMILUMINESCENCE

Enrique Cadenas

Department of Pathology II
Linköping University
Linköping, Sweden

LOW-LEVEL CHEMILUMINESCENCE. PERSPECTIVES

The whole process of chemiluminescence can be considered as the formation of an excited product (P*) from starting reactants A and B, followed by transition of the excited molecule P* to the ground state P with emission of a photon:

$$A + B \rightarrow P^* \rightarrow P + h\upsilon \qquad (1)$$

This is referred to as *low-level-*, *ultraweak-*, or *dark-chemiluminescence*. The occurrence of excited states in biological reactions can be considered as including two basic processes (Fig. 1.): (I) the formation of an electronically excited state and (II) the decay of the excited state to the ground state. The former could originate from a biochemical pathway involving, among others, interaction of free radicals (e.g., lipid peroxidation), oxidation of several quinones, and enzymatic reactions which yield excited oxygen or carbonyl compounds. The latter can occur through different routes, such as photon emission (giving rise to low-level chemiluminescence), physical quenching, energy transfer to adequate acceptors, or reaction with different bioconstituents. The excitation energy of these excited states could be utilized to drive photoprocesses in the dark (Cilento, 1982).

Other expressions of excited state formation in biological systems which concern the emission arising from cell growth (Quickenden et al., 1985) or the application of the low-level chemiluminescence technique in the fields of medicine and immunology (Allen, 1980, 1982; Allen & Lieberman, 1984; Gisler et al., 1982, 1983) are not included in this survey. The reaction of $O_2^-\cdot$ with luminol, which might have some diagnostic values in biological and immunological systems (also in phagocytosis) was recently studied in terms of mechanism and quantification of chemiluminescence efficiency (Merenyi et al., 1985; Dahlgren & Briheim, 1985).

117

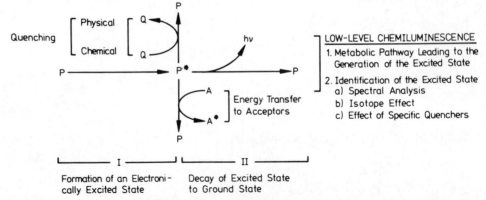

Figure 1. Basic processes involving electronically excited states. The generation of the excited states during a biochemical reaction (I) is followed by their deactivation (II) through different competitive processes: (a) emission of a photon, (b) physical quenching, (c) chemical reaction, and (d) energy transfer.

EXCITED STATES: SINGLET OXYGEN AND EXCITED CARBONYLS

An electronically excited state results from the promotion of an electron to higher energy orbitals. In almost all molecules, the electrons are paired with electrons of opposite spin. A molecule with no net unpaired electrons is in the singlet state (S_0) (Fig. 2). Promotion of an electron to a higher orbital without change in the spin results in the first singlet excited state (S_1). The singlet can, in some cases, undergo spin conversion (intersystem crossing = ISC) to give a triplet state (T_1), which has two unpaired electrons. Two radiative (emission of light) processes originate from the decay of the lowest singlet or triplet to the ground state. Radiative transitions between states of like multiplicity ($S_1 \rightarrow S_0$; fluorescence) have a rate constant $k_f = 10^7–10^9$ s^{-1}, whereas radiative transitions between states of unlike multiplicity ($T_1 \rightarrow S_0$; phosphorescence) have a relatively longer half-life ($k_p = 10–10^4$ s^{-1}).

Energy transfer from the excited state to an appropriate acceptor in the cell would be feasible as long as Forster (1959) criteria are met, that is the requirement of close proximity of the acceptor to the emitter at the moment of its generation, such that overlap of orbitals is possible. This type of energy transfer might play an important role in biological chemiluminescence reactions.

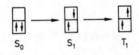

Figure 2. Schematic representation for the formation of a singlet and triplet excited state.

Molecular oxygen, a ground state triplet with paramagnetic and diradical-like properties, has two electrons in separate orbitals ($\pi_{2px}^{*1} = \pi_{2py}^{1*}$) parallel spins (simplified in Table 1); this is designated as $^3\Sigma g-$ (Σ is a symbol signifying a state with axial symmetry, and the superscript 3 means triplet). The first excited state is a singlet designated $^1\Delta g^-$ (both electrons in a single orbital with opposed spins). Its energy lies 22 kcal above the ground state; as the transition $^1\Delta g \rightarrow {}^3\Sigma g^-$ is spin forbidden, it is a relatively long-lived species, and its lifetime in H_2O is about 1×10^{-6} s. The second singlet excited state ($^1\Sigma g^+$) has two electrons with opposite spin but in different orbitals: It is 37.5 kcal above the ground state and has a much shorter lifetime (10^{-11} s) and decays to the $^1\Delta g$ state. The lifetime of $^1\Sigma g^+$ is sufficiently short that all singlet oxygen chemistry in solution involves the $^1\Delta g$ state. There is no evidence for the occurrence of this second excited state in biological systems. The term singlet oxygen as $^1O_2$ will be used in this context to designate the $^1\Delta g$ state of oxygen.

Table 1. Electronic states of molecular oxygen.

State		Orbital occupancy	Energy above ground state kcal	Life time s
Ground	$^3\Sigma$			
First excited	$^1\Delta$		22.5	10^{-6}
Second excited	$^1\Sigma$		37.5	10^{-11}

The molecular electronic transitions occurring for the case of an excited carbonyl compound might imply either a $^3n,\pi^*$ or $^3\pi,\pi^*$ state. The lowest energy transitions for molecules with nonbonding electron pairs is of n,π^* character and it is common for O_2-containing compounds. In the n,π^* transition, an electron is removed from the carbonyl oxygen lone pair and promoted to a vacant π^* orbital shared by carbon and oxygen. An excited carbonyl could be generated in the triplet state and, because of its longer lifetime (in comparison with the singlet state), it could be expected that it would have a wider relevance in biological reactions by reacting with or transferring energy to biomolecules. As a matter of fact, a large body of photochemistry is found to originate from the triplet state.

DEACTIVATION OF THE EXCITED STATE

Photon Emission

Singlet oxygen. Decay to the ground state of $^1\Delta g$ oxygen is associated with light emission peaking at 1270 nm (k = 7900 cm^{-1}). This single molecule transition with emission of light is known as singlet oxygen monomol emission (reaction 2).

$$^1\Delta O_2 \rightarrow {}^3\Sigma O_2 + h\upsilon \ (1270nm) \tag{2}$$

The higher energy state shown by emission at 633 nm is formed by interaction of two excited O_2 molecules (a one-photon two-molecule collisionally-induced process), the energy of the excited state arising from a non resonance, superposition of two excited states energies of the component species. These are known as simultaneous transitions of pairs. The 633 nm emission originates from the energy of two $^1\Delta$ molecules and the 703 nm emission arises from the same $^1\Delta$ complex when one of the O_2 molecules ends up in the first vibrational level (Khan & Kasha, 1970).

$$2 \ ^1\Delta O_2(O,O) \rightarrow 2 \ ^3\Sigma O_2 + h\upsilon \ (634 \ nm) \tag{3}$$

$$2 \ ^1\Delta O_2(O,1) \rightarrow 2 \ ^3\Sigma O_2 + h\upsilon \ (703 \ nm) \tag{4}$$

Figure 3 illustrates the singlet oxygen dimol emission arising from the peroxide-hypohalite reaction (Seliger 1960; Khan & Kasha, 1970).

$$H_2O_2 + OCl^- \rightarrow H_2O + Cl^- + {}^1O_2 \tag{5}$$

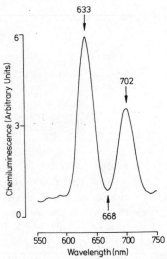

Figure 3. Emission spectrum of singlet oxygen generated during the OCl^-/H_2O_2 reaction. The spectrum was taken with a spectrofluorimeter (Perkin-Elmer, model LS-5). OClNa (0.8 M) was pumped into a 1.6 M H_2O_2 solution at a rate of 220 mol/min.

Excited carbonyls. The weak emission arising from triplet or singlet carbonyls on decaying to the ground state takes place around 380-450 nm (reaction 6).

$$>C{=}O^* \rightarrow >C{=}O + h\upsilon \ (380\text{--}450 \ nm) \tag{6}$$

This excitation can be emitted directly or transferred to a fluorescer. In some systems the formation of these triplet species might be reasonably postulated to occur, though the presence of O_2 (an efficient quencher of triplet species with a cellular concentration of about 10^{-4} M) makes the identification of triplet carbonyls difficult. An approach for the study of excited triplet carbonyl compounds is based on transfer of their energy to adequate acceptors (A) (e.g., 9,10-dibromoanthracene, DBA), thus populating the fluorescence state (^{1}A) of the acceptor (Figure 4). Triplet-singlet energy transfer would be the dominant mechanism in the case of "heavy atom" substituted anthracenes (e.g., DBA), whereas singlet-singlet energy transfer is dominant in the case of "nonheavy atom" substituted anthracenes (e.g., DPA). Thus, the triplet nature of a carbonyl compound could be further attested by the failure to excite acceptors which lack the "heavy" atom necessary for spin exchange (Cilento, 1980).

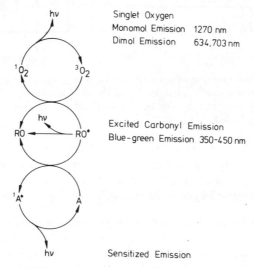

Figure 4. Energy transfer from an excited triplet carbonyl to oxygen and a fluorescence acceptor. The excitation of a triplet carbonyl (RO*) can be emitted directly or transferred either to O_2 or to a fluorescer (A).

Detection of chemiluminescence. The single-photon-counting apparatus in Figure 5 uses a red-sensitive photomultiplier cooled to -25 °C by a thermoelectric cooler (EMI Gencom, Plainview, N.Y.) in order to reduce the dark current. Suitable photomultipliers with S-20 response are EMI 9658AM, RCA4832, and Hamamatsu HTV R374. The photomultiplier is connected to an amplifier-discriminator (EG&G Princeton Applied Res., Princeton, N.J.) adjusted for single-photon counting. This, in turn, is connected to a frequency counter and/or a chart recorder. Efficient light-gathering from the sample is established by using a lucite rod as optical coupler or an ellipsoidal light reflector. A shutter allows a continuous operation of the photomultiplier in order to remain dark adapted. A thermostated cuvette for cellular or subcellular and enzyme suspensions is placed inside a light-tight box and additions are made from the outside using precision micropumps and light-shielded tubing. The thermostated cuvette is equipped with a magnetic stirrer to maintain homogeneous suspension, and the lid contains ports for an O_2 electrode and for required tubings, e.g., for injection of compounds or for gas exchange. Responses to additions are often in the form of a burst of emission; therefore additions are made through tubing from outside of the light-tight box, in order to avoid intervals without recording.

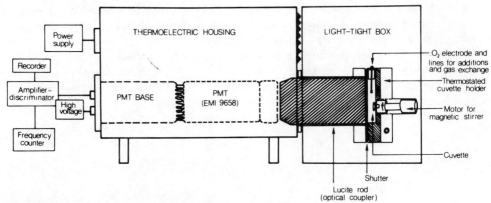

Figure 5. Single photon-counting apparatus for the measurement of low-level chemilumines-cence. This setup illustrates the "cuvette mode" of the photon counter. Further applications for chemiluminescence measurements *in vivo* and of perfused organs (non-invasive technique) are described in Boveris et al., 1981.

For spectral analysis of emitted light, optical filters may be placed into the light path. These can be interference filters or cut-off filters consisting of colored glass or Wratten gelatin filters. Filters may be in a circular array (with automatic insertion through a filter drive controller) or exchangeable by slits.

In addition, this setup can be used for chemiluminescence measurements of exposed organs *in situ* or isolated perfused organs (Boveris et al., 1981; Cadenas & Sies, 1984). Based on its noninvasive character, low-level chemiluminescence was recently utilized to detect pathological conditions in biological samples. Observations on spontaneous chem-iluminescence of blood plasma and urine of normal and cancer-bearing patients consti-tute the first examples in this direction (Gisler et al., 1982, 1983).

Reaction with Cellular Constituents

Low-level chemiluminescence by living systems originates from the biochemical gen-eration of electronically excited states. The excited singlet and/or triplet species may be formed in the cellular milieu in high yields, but unless they have a chance to radiate they will go undetected; indeed these excited states may be almost totally quenched by undergoing cellular physical and chemical processes.

For example, several aminoacids (present within the cells in mM range) and pro-teins (Matheson et al., 1975), vitamin C (Rooney, 1983; Chou & Khan, 1983), vitamin E (Grams & Eskins, 1972; Foote, 1976), vitamin A (Smith, 1983), NADPH, NADH, and certain components of the mitochondrial electron transport chain (Bodaness, 1982), among others, are effective quenchers of $^1O_2$. Some examples are illustrated in Table 2. On the other hand, several biologically-relevant compounds, such as indoles, tyrosine and its 3,5-dihalogeno-derivatives, quinones, riboflavine, xanthene dyes, etc., are known to quench triplet carbonyls (e.g., chemiexcited triplet acetone) in aqueous medium (see Catalani & Bechara, 1984). This shows that cells are endowed with a variety of physical and chemical quenchers of excited species, making spontaneous chemiluminescence of liv-ing tissues (Barenboim et al., 1969; Cadenas et al., 1984) hardly detectable.

Table 2. Rate constants for the interaction of $^1O_2$ with several compounds of biological relevance. k represents the overall rate constant unless k_r (chemical reaction rate constant) or k_q (quenching rate constant) is specified.

Compound	k (M^{-1} s^{-1})	Reference
1. Aminoacids and proteins		
Histidine	3.2×10^7	Kraljic and Sharpatyi, 1978; Matheson et al., 1975
	10.0×10^7	Matheson and Lee, 1979
Tryptophan	1.0×10^7	Nilson et al., 1972
	3.0×10^7	Smith, 1978; Matheson and Lee, 1979
Methionine	1.0×10^7	Nilsson et al., 1972
	1.7×10^7	Matheson and Lee, 1979
Tyrosine	0.8×10^7	" " " "
	2.7×10^7	Weil, 1965
Alanine	0.2×10^7	Matheson and Lee, 1979
Arginine	1.0×10^6	Kraljic and Sharpatyi, 1978
Histamine	2.8×10^7	" " " "
SOD	2.6×10^9	Matheson et al., 1975
Apo-SOD	1.5×10^9	" " " "
Carbonic anhydrase	4.5×10^8	" " " "
Lysozime	1.3×10^8	Kepka and Grossweiner, 1973
Trypsin	7.1×10^9	Stevens, 1973
2. Vitamins and Coenzymes		
Vitamin C	8.3×10^6	Bodaness and Chan, 1979; Chou and Khan, 1983
ß-carotene	3.0×10^{10}	Foote and Denny, 1968
Vitamin E	1.0×10^8	Grams and Eskins, 1972
	2.5×10^8	Fahrenholtz et al., 1974
	4.6×10^7 (k_r)	Foote et al., 1974
	6.7×10^8	" " "
Ergosterol	1.3×10^7	Shenck et al., 1957
Lipoic acid	1.0×10^8	Stevens, 1973
NADPH	3.0×10^8	Bodaness and Chan, 1977
Cytochrome c	4.9×10^8	Peters and Rodgers, 1980
3. Lipids		
Methyl oleate	7.4×10^4	Krasnovsky et al., 1983
Methyl linoleate	1.3×10^5	" " " "
	1.2×10^5 (k_r)	Doleiden et al., 1974
Methyl linolenate	1.9×10^5	Krasnovsky et al., 1983
	1.9×10^5 (k_r)	Doleiden et al., 1974
Methyl arachidonate	2.6×10^5 (k_r)	" " " "
Cholesterol	6.6×10^4	Shenck et al., 1957
Oleate	1.7×10^4 (k_q)	Krasnovsky et al., 1983
Linoleate	4.2×10^4 (k_q)	" " " "
Linolenate	8.0×10^4 (k_q)	" " " "
Arachidonate	10.0×10^4 (k_q)	" " " "
Palmitic	8.0×10^3 (k_q)	" " " "
Stearic	9.0×10^3 (k_q)	" " " "
4. Nucleosides - Nucleotides		
Adenine	1.0×10^6	Kraljic and Scharpatyi, 1978
Thymine	1.0×10^6	" " " "
Cytidine	1.0×10^6	" " " "
Adenosine	1.0×10^6	" " " "
Guanosine	1.0×10^6	" " " "
5. Miscellaneous compounds		
Bilirrubin	3.5×10^8 (k_r)	Matheson, 1979
	9.0×10^8 (k_q)	" "
α-naphtol	3.2×10^7	Snyakin et al., 1978
β-naphtol	7.6×10^6	" " "
Hydroquinone	7.0×10^7	Foote et al., 1970
Benzoquinone	3.4×10^7	Koka and Song, 1978

Ascorbic acid reacts with $^1O_2$ (Chou & Khan, 1983), as indicated by its quenching of $^1O_2$ emission at 1280 nm. The quenching of $^1O_2$ by vitamin C seems to be predominantly chemical, and corroborates previous studies on the reactivity between $^1O_2$ and ascorbate (Bodaness & Chan, 1979). The quenching rate of triplet carbonyl compounds (triplet benzophenone) by ascorbic acid has a high value ($1.2 \times 10^9 \text{ M}^{-1}\text{s}^{-1}$), indicating that the carbonyl triplet promotes efficient H abstraction from the HO group of vitamin C (Encinas et al., 1985).

Vitamin E, in addition to its known interaction with free radicals [also including $O_2^-\cdot$ (Ozawa et al., 1983)], which gives it the property of an efficent lipid-soluble, chain-breaking antioxidant (Burton et al., 1983) was observed to react with $^1O_2$ with a rate constant of about $10^8 \text{ M}^{-1}\text{s}^{-1}$. The reaction of vitamin E with $^1O_2$ involves both oxidation of the phenol and quenching of $^1O_2$ by the phenol without reaction. The reactivity of α-tocopherol towards $^1O_2$ was the highest compared with other tocopherols (Grams & Eskins, 1972; Foote, 1976) and it is only about 50-fold slower than that of the diffusion-controlled quencher, β-carotene. The high reactivity of vitamin E towards triplet carbonyl compounds ($6.7 \times 10^9 \text{ M}^{-1}\text{s}^{-1}$; Encinas et al., 1985) parallels the particularly large reactivity of vitamin E towards O_2-centered radicals (Burton et al., 1983). Quenching of a triplet carbonyl could be explained — similarly to vitamin C — in terms of a H abstraction from the OH group of the tocopherol. The values above were obtained from model systems, and the likelihood of such reactions within biological membranes remains to be established. However, it could be speculated that the ability of vitamin E to protect against oxidative damage relies on its quenching capacity towards O_2-centered radicals, $^1O_2$, and triplet carbonyl compounds.

The physical quenching of $^1O_2$ by β-carotene is long known (Foote & Denny, 1968). The capacity of β-carotene of reaction with $^1O_2$ establishes the basis for its antioxidant activity in biological systems, protection on cellular damage caused by visible light, and treatment of certain photosensitive diseases. The mechanism of carotenoid protection, however, does not rely only on its physical quenching of $^1O_2$, but also on its quenching capacity of triplet sensitizers and inhibition of free radical reactions (Krinsky, 1982).

Aminoacids react rapidly with photophysically-generated $^1O_2$, the interaction being largely chemical. Histidine, tryptophan, and methionine quench $^1O_2$ with rate constants of about $10^7 \text{ M}^{-1}\text{s}^{-1}$ estimated by direct photophysical methods and via methylene blue photosensitization (Matheson et al., 1975; Nilsson et al., 1972). The quenching effect of proteins containing these aminoacids is similar to those of the aminoacids free in solution. Matheson et al. (1975) speculate that quenching of $^1O_2$ in biological systems would dominantly proceed through aminoacids, given their cellular concentration and their overall quenching rates with $^1O_2$ (10^6 s^{-1}). In the case of the aminoacids quoted above, there is no evidence for physical quenching in addition to the chemical reaction (Matheson & Lee, 1979). The high reactivity of the SH group of tripeptide GSH towards free radicals could account, in a joint activity with vitamin E, for the temporary protection against lipid peroxidation, detected by low-level chemiluminescence (Bartoli et al., 1983). Moreover, GSH quenches efficiently triplet carbonyls ($6.7 \times 10^8 \text{ M}^{-1}\text{s}^{-1}$; Encinas et al., 1985) by a reaction involving its -SH group. The quenching process seems to proceed by a charge transfer complex followed by H abstraction of the S-H bond and/or the α-carbon atom (Inbar et al., 1982). Other thiol-containing compounds, such as cystein, methionine, and GSSG, also quench triplet carbonyls efficiently (k_Q = 3.2–, 3.5–, and 13 × 10^8 $M^{-1}s^{-1}$, respectively). The high k_Q value of GSH towards triplet carbonyls, along with the large concentration of the former in the cellular milieu, makes GSH an efficient defense against possible damage by excited carbonyl species.

Some other cell constituents of the cytosol might be regarded as exerting a protective role: NADPH and NADH were found, upon their reaction with $^1O_2$ to protect enzymes from $^1O_2$ deactivation. The reaction of $^1O_2$ with NADPH was estimated to proceed via a 2-electron transfer process with a rate constant of 3 × 10^8 $M^{-1}s^{-1}$ (Bodaness & Chan, 1977). The reaction could also proceed through an intermediate step with formation of NADP·; this step is not facile to demonstrate because of the faster rate with which O_2 subsequently reacts with NADP· (Peters & Rodgers, 1980). After its reaction with $^1O_2$, NADPH could be regenerated by the concerted action of the cytosolic enzymes responsible for maintaining a constant intracellular $NADP^+$/NADPH ratio (Bodaness & Chan; Bodaness, 1982). Cytochrome c reacts with $^1O_2$ at a rate of 4.9 × 10^8 $M^{-1}s^{-1}$ (Peters and Rodgers, 1980).

Saturated fatty acids seem to quench $^1O_2$ via a physical mechanism, whereas the quenching by unsaturated fatty acids seems to proceed by a chemical mechanism (with quenching rate constants ranging between 2 and 15 × 10^4 $M^{-1}s^{-1}$ (Krasnovsky et al., 1983). The allylic and double allylic bonds of unsaturated fatty acids would exert the highest quenching activity of $^1O_2$. It was reported that a good correlation for the rate constant for reaction of $^1O_2$ with several fatty acids is found when a partial rate of 0.19 × 10^5 $M^{-1}s^{-1}$ and 0.28 × 10^5 $M^{-1}s^{-1}$ is assigned for each allylic and double allylic hydrogen, respectively (Doleiden et al., 1974).

These interactions with cellular constituents (Table 2) can be accounted for by physical quenching and/or chemical reaction. Three main types of $^1O_2$ reactions with alkanes — which will participate to a different extent in the oxidation of biological substrates — are well known: an "ene"-type reaction with olefines to yield allylic hydroperoxides, 1,4-cycloaddition to diene systems to give cyclic peroxides, and 1,2-cycloaddition to olefins to form 1,2-dioxetanes (Foote, 1976). On the other hand, excited ketones are capable of H abstraction from suitable donors, for example polyunsaturated fatty acids, with formation of free radical products which can initiate free radical chain processes. In this regard, it is worth noting that the behavior of the n,π* carbonyl triplets resembles that of alkoxy radicals (Wagner, 1971; Encinas et al., 1981).

The high reactivities of GSH and vitamins E and C towards triplet carbonyls makes them relevant defenses against the cellular damage exerted by potentially excited carbonyl-generating reactions, such as membrane lipid peroxidation. However, it must be considered that the interactions of excited species, such as triplet carbonyls, with cellular constituents might result in useful and/or detrimental processes in biological systems; in this regard, several photochemical-like transformations occur *in vivo* in the absence of light (White & Wei, 1970; White et al., 1974).

BIOLOGICAL GENERATION OF EXCITED STATES

The weak chemiluminescence from biological systems may conceivably have various origins and it exhibits an intricate spectral distribution, thus revealing that light emission might be more complex than that observed in the relaxation of a sole excited state to the ground state.

The participation of mainly two excited states in biological chemiluminescence is surveyed in this chapter, singlet molecular oxygen and excited carbonyl compounds. The former species can originate from a hydroperoxide in reactions implying a tetroxide (Howard & Ingold, 1968) or a zwitterion (Bowman et al., 1971) intermediate, recombination of O_2-containing radicals (Krinsky, 1979), and energy transfer from a triplet ketone to triplet O_2 (Wu & Trozzolo, 1979). The latter species can originate from an intermediate dioxetane (as postulated for several peroxidase-catalyzed reactions; Cilento, 1980), an electron transfer process, either enzymatic (Villablanca & Cilento, 1985) or non-enzymatic (Ginsberg & Cadenas, 1985), and free radical recombination reactions. The recombination of peroxy radicals proceeds via a tetroxide intermediate with elimination of a ketone, an alcohol, and molecular O_2. This mechanism was reported to proceed via the elimination of $^1O_2$ (Howard & Ingold, 1968) or a ketone in the triplet state (Kellog, 1969). The occurrence of excited carbonyl luminescence involving a dioxetane precursor, on the other hand, might imply the involvement of $^1O_2$: the reaction of the latter with electron-rich double bonds to yield dioxetanes proceeds very efficiently and photoemission arising from the dioxetane scission would follow.

The above-quoted chemical reactions are summarized in Figure 6, and they might serve as models for reactions occurring in biological systems and contribute to the phenomenon of biological chemiluminescence: they would be involved to different extent in the photoemission accompanying lipid peroxidation, redox reactions involving quinones, and certain enzymatic (peroxidase-catalyzed) reactions described below.

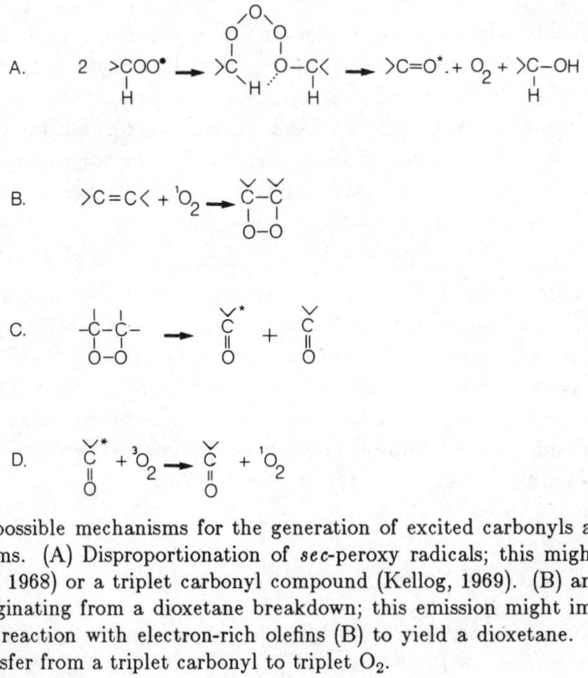

Figure 6. Some possible mechanisms for the generation of excited carbonyls and singlet oxygen in biological systems. (A) Disproportionation of *sec*-peroxy radicals; this might yield either $^1O_2$ (Howard & Ingold, 1968) or a triplet carbonyl compound (Kellog, 1969). (B) and (C) Triplet carbonyl emission originating from a dioxetane breakdown; this emission might imply the participation of $^1O_2$ by its reaction with electron-rich olefins (B) to yield a dioxetane. (D) Generation of $^1O_2$ by energy transfer from a triplet carbonyl to triplet O_2.

Detailed discussions covering several other aspects of biological chemiluminescence have been provided (Krinsky, 1979; Vladimirov et al., 1980; Allen, 1982; Duran, 1982; Slawinska & Slawinski, 1983; Cadenas, 1985).

Lipid Peroxidation

Studies on the photoemission associated with lipid peroxidation show that the excited species thereby formed are related to the interaction of lipid-derived radicals formed in the propagation steps of lipid peroxidation (Nakano & Noguchi, 1977). These studies confirmed spectroscopically the generation of $^1O_2$ in the termination step of microsomal lipid peroxidation, as a consequence of the self-reaction of lipid peroxy radicals (Russell, 1957; Howard & Ingold, 1968). The same mechanism may alternatively lead to the formation of a carbonyl compound in the excited triplet state (Kellog, 1969) (Fig. 6A). The formation of a dioxetane compound, after reaction of $^1O_2$ with unsaturated fatty acids (Fig. 6B), may account for an additional source of excited carbonyl groups through dioxetane breakdown (Fig. 6C). Another independent posibility for the generation of $^1O_2$ during peroxidation would be the quenching of triplet carbonyls by O_2 (Fig. 6D) (Wu & Trozzolo, 1979). Therefore, it might be concluded that the existence of $^1O_2$ could also imply the existence of triplet carbonyls.

The formation of carbonyl compounds during lipid peroxidation is well known (Esterbauer, 1982). Some of these non-radical products of lipid peroxidation, like 4-hydroxy-2,3-trans-nonenal exert irreversible cellular damage upon incubation with isolated hepatocytes, reflected as GSH depletion and enhanced O_2-induced chemiluminescence and formation of volatile hydrocarbons (Cadenas et al., 1983b).

None of these electronically excited molecules, however, seem to account for the total light emission observed during lipid peroxidation, and quantitative studies are still required. The molecular mechanism summarized above would be operative in different cellular oxidative situations associated with lipid peroxidation and have been inferred (to a different extent) in different biological chemiluminescence models: (a) rat liver microsomes supplemented with NADPH (Nakano & Noguchi, 1977), ascorbate (Cadenas et al., 1984b), or organic hydroperoxides (Cadenas & Sies, 1982; Cadenas et al., 1983c), (b) isolated hepatocytes under hyperoxic conditions (Cadenas et al., 1981b), (c) tissue homogenates supplemented with lipid (Miyazawa et al., 1981) or organic (Cadenas et al., 1981a) hydroperoxides, under hyperoxic conditions (Cadenas et al., 1981a), or supplemented with CCl_4 (Miyazawa et al., 1984), and (d) intact animals fed a diet containing lipid hydroperoxides (Miyazawa et al., 1983a,b,c), during chronic alcoholism treatment (Boveris et al., 1983), or dosed with CCl_4 (Miyazawa et al., 1984).

On the other hand, hemoprotein/organic hydroperoxide interactions were shown to be associated with the formation of excited states (Slawinski et al., 1981; Cadenas, et al., 1980a,b,c), a process which might contribute to biological chemiluminescence. The lipid hydroperoxide/microsomal cytochrome P-450 interaction provides the base for the lipid hydroperoxide-dependent initiation of lipid peroxidation (Aust & Svingen, 1982). The reaction between organic hydroperoxide and cytochrome P-450 is associated with low-level chemiluminescence, and the generation of excited states can be accounted for by the cytochrome-catalyzed heterolytic (Ullrich, 1977) or homolytic (McCarthy & White,

1983) scission of organic hydroperoxides. The former mechanism implies the reaction of hydroperoxide with cytochrome P-450 with formation of an intermediate oxo-ferryl complex which, upon reaction with a second molecule of hydroperoxide, yields O_2 in the singlet state. This mechanism involves therefore hydroperoxide disproportionation reactions and is directly linked to the generation of excited species (Cadenas et al., 1983c). Light emission can be observed under anaerobic conditions when a heterolytic cleavage of the hydroperoxide is operative, thus providing evidence that the photoemissive species originate from the oxygen atom of the hydroperoxide. During the homolytic scission of hydroperoxide, however, the excited species would be only generated in a secondary fashion, as a consequence of lipid peroxidation triggered by oxyradicals formed during hydroperoxide cleavage (Cadenas & Sies, 1982).

Independent of the nature of the excited states formed during the hydroperoxide-induced chemiluminescence of microsomal fractions, the occurrence of photoemission requires the absence of reducing equivalents to cytochrome P-450 (which are ordinarily used to initiate lipid peroxidation by iron reduction) and of monooxygenase substrates.

The presence of excited triplet carbonyls in lipid peroxidation processes was suggested on the basis of indirect evidence or a weak emission in the blue-green spectral region. The use of chlorophyll-a, which is an efficient detector of triplet carbonyls (Brunetti et al., 1983; Nassi & Cilento, 1983), permitted the identification of triplet carbonyls during photoemission arising from t-butyl hydroperoxide-supplemented microsomal membranes (Cadenas et al., 1984c). This intense photoemission suggests that triplet excited carbonyl compounds are generated substantially during the process of lipid peroxidation and, moreover, that their formation does not occur solely in termination steps.

The aerobic oxidation of fatty acids containing a 1,4-*cis,cis*-pentadiene system by lipoxygenase yields excited states (Boveris et al., 1980; Schulte-Herbrüggen & Cadenas, 1985a,b). This process can be considered as an enzymatic lipid peroxidation and the formation of excited states as dependent on the presence of free radical flow, which accounts for about 2-4% of the total enzymatic activity. Both the intensity of light emission and O_2 consumption are linearly related to the number of double bonds present in the fatty acid substrate: linoleate (18:2) > linolenate (18:3) > arachidonate (20:4). Oleate (18:1) elicits neither photoemission nor O_2 consumption, since it lacks the 1,4-*cis,cis*-pentadiene system required for the enzyme activity.

The lipoxygenase-catalyzed oxidation of arachidonate generates excited species with an efficiency of 4.5×10^{-10} photons/O_2 molecules. The occurrence of triplet carbonyls during this reaction is attested by the excitation of different acceptors (See Fig. 4), thus eliciting sensitized emission. DBAS- and chlorophyll-a-sensitized emission of the arachidonate/lipoxygenase reaction proceed with an efficiency 4.5×10^3 higher than in the absence of sensitizers. Since both acceptors have different energy requirements to populate their fluorescence state, two possible mechanisms for the generation of triplet carbonyls could be postulated. (I) Free radical interactions yielding conjugated carbonyl compounds with energy enough to promote chlorophyll-a-sensitized emission, and (II) breakdown of a possible dioxetane intermediate from lipid hydroperoxides yielding unconjugated carbonyls, with energy enough to promote both DBAS- and chlorophyll-a-sensitized emission (Schulte-Herbrüggen and Cadenas, 1985a). These possibilities are summarized in Figure 7. The complex photoemission from mechanically-injured soybean root tissue cannot be ascribed to the lipoxygenase activity — present in high amounts in this tissue — but rather to other enzymatic systems not yet identified (Salin et al., 1985).

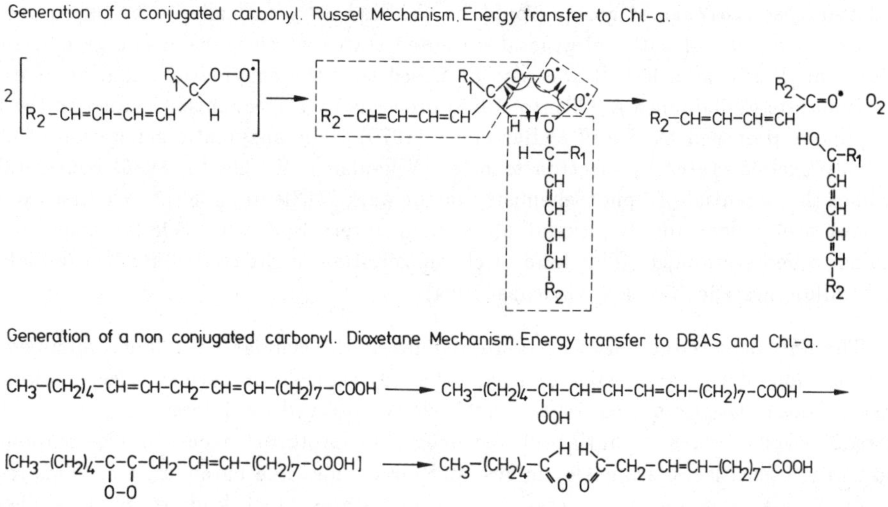

Generation of a conjugated carbonyl. Russel Mechanism. Energy transfer to Chl-a.

Generation of a non conjugated carbonyl. Dioxetane Mechanism. Energy transfer to DBAS and Chl-a.

Figure 7. Two mechanisms to account for the generation of triplet carbonyls during lipid peroxidation and the occurrence of sensitized emission.

Chemiluminescence During Redox Transitions of Quinones

Several redox transitions involving quinones have been shown to be accompanied by generation of excited states. Examples are given by the chemiluminescence observed during oxidation of adrenaline, adrenochrome (Slawinska, 1978) and other quinones (Slawinska & Slawinski, 1973), the autoxidation of 6-hydroxydopamine (Heikkila & Cabbat, 1978), the one-electron activation of paraquat (Cadenas et al., 1983a) and menadione (Wefers & Sies, 1983), the iron/H_2O_2-mediated activation of bleomycin (Trush et al., 1983) the nonenzymatic (presumably H_2O_2/O_2^- mediated) oxidation of p-benzoquinone (Ginsberg & Cadenas, 1985) and the enzymatic oxidation of cathecol (Villablanca & Cilento, 1985).

Although the mechanism supporting the formation of electronically excited states is different from the one supporting lipid peroxidation, the examples quoted above seem not to share a unique mechanism. The chemiluminescence observed during these redox transitions was ascribed in some cases to the generation of excited forms of the quinone and, in others, to the generation of $^1O_2$.

The chemical oxidation of quinones or semiquinones leads to the generation of a triplet semiquinone (reaction 7) or a triplet quinone (reaction 8) (Stauff & Bartolmes, 1970), which upon decay emit 515 and 568 nm, respectively.

$$\text{HQ} \rightarrow {}^3\text{SQ*} \rightarrow \text{SQ} + h\upsilon \tag{7}$$

$$\text{SQ} \rightarrow {}^3\text{Q*} \rightarrow \text{Q} + h\upsilon \tag{8}$$

This observation seems to apply to the chemiluminescence observed during the oxidation of p-benzoquinone (Ginsberg & Cadenas, 1985) and cathecol (Villablanca et al., 1985). In the former case, oxidation would be promoted nonenzymatically by O_2^-/H_2O_2, whereas in the latter case, an enediol group would be oxidized enzymatically to a diketo group; moreover, oxidation of cathecol generates excited states which transfer energy efficiently to chlorophyll-a in micelles. It was also proposed that the enzymatic oxidation of dihydroxyfumarate (Villablanca & Cilento, 1985) proceeds by a mechanism leading to the triplet quinone proposed by Stauff & Bartolmes (1970). The enzymatic generation of electronically excited states by electron transfer (Villablanca & Cilento, 1985) considerably amplifies the potential of "photochemistry in the dark" (Cilento, 1982). Whether excited quinones are biologically functional upon their interaction with cellular constituents remains to be determined. This kind of chemiexcitation might contribute to the biological chemiluminescence process (Cadenas, 1984).

The univalent reduction of certain quinones at the cellular level is accompanied by the formation of O_2^- and, subsequently, other O_2 radicals or derived species (Borg & Schaich, 1984). O_2 is required for the cytotoxicity observed by different xenobiotics during redox cycling, such as antineoplastic drugs, antiprotozoic agents, nitro-compounds used as radiosensitizers, aromatic amines, and insecticides like paraquat, etc. This redox cycling process promotes the generation of excited states which leads to low-level chemiluminescence (Fig. 8). The redox cycling-supported light emission is sensitive to superoxide dismutase, thus pointing to the relevance of O_2^- in the chemiluminescence observed during redox cycling of paraquat (Cadenas et al., 1983a) and menadione (Wefers & Sies, 1983).

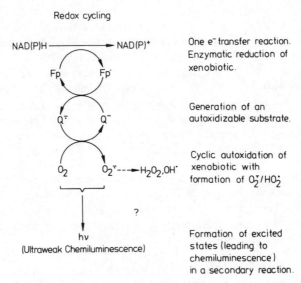

Figure 8. One-electron activation of quinone compounds and the generation of low-level chemiluminescence.

The molecular mechanism for the generation of excited species are apparently different from those described above, and seem to be multifaceted. This type of superoxide dismutase-sensitive photoemission can be partly explained as originating from the spontaneous disproportionation of $O_2^{-\cdot}$ (reaction 9), known to produce $^1O_2$.

$$O_2^{\cdot-} + HO_2^{\cdot} + H^+ \rightarrow H_2O_2 + {}^1O_2 \tag{9}$$

Although theoretical and experimental evidence is available to support this reaction (Stauff et al., 1963; Khan, 1981; Koppenol, 1976), evidence against this has been brought forward (Foote et al., 1981), and its likelihood in biological systems remains under debate.

Moreover, red light emission originating during the oxidation reactions quoted above does not account for the total light emission and, for the case of paraquat-induced chemiluminescence of microsomal fractions, it is feasible that the diradical pair anihilation between $O_2^{-\cdot}$ and a semiquinone cation radical ($SQ^{\cdot+}$) could yield an excited state of the acceptor and molecular O_2 in the ground state (reaction 10) (Faulkner & Glass, 1982).

$$O_2^{\cdot-} + SQ^{\cdot+} \rightarrow {}^3O_2 + {}^1SQ^* \tag{10}$$

Another possibility might be given by the radical dehydrogenation of a biological substrate (promoted by a Fenton-type reaction involving $O_2^{-\cdot}$), producing a radical product substrate. The reaction of the latter with O_2 forms a peroxide or dioxetane-like product, which, upon breakdown, may account for the formation of excited species, probably an excited triplet carbonyl (Allen, 1982). This last possibility, along with the mechanisms illustrated in reactions 9 and 10 are dependent on the presence of $O_2^{-\cdot}$.

The occurrence of $O_2^{-\cdot}$ is the first step in a complex series of redox reactions including radical and non radical oxidants and photoemission. The inhibition by superoxide dismutase of redox cycling-supported photoemission therefore only suggests the participation of $O_2^{-\cdot}$ in some stage of the process leading to light emission, but does not point out to a unique-prevailing mechanism for the generation of electronically excited molecules.

Another type of chemiluminescence is also present in metabolic states conducive to the mitochondrial formation of $O_2^{-\cdot}$ and H_2O_2 (Boh et al., 1982), the generation of which is known to occur under certain conditions at specific sites of the mitochondrial electron transport chain (Boveris & Cadenas, 1982). This photoemission is based on the ability of acetaldehyde to react with $O_2^{-\cdot}$ or H_2O_2 generated by mitochondria to form a metastable intermediate which decays spontaneously with emission of light. It was proposed that a triplet carbonyl function and/or $^1O_2$ could account for this photoemission (Boh et al., 1982).

Enzymatic Generation of Excited Carbonyls

Triplet carbonyl species are generated in high yields by the peroxidase-catalyzed aerobic oxidation of adequate substrates (Cilento, 1984); this process might proceed through the formation of a hypothetical dioxetane (Fig. 6C). These enzyme-generated triplet carbonyl species are able to (a) transfer energy to appropriate acceptors (thus

131

eliciting sensitized emission) and (b) promote photochemical processes. For the case of peroxidase-catalyzed generation of triplet acetone, the former process was observed as energy transfer to dibromoanthracene sulfonate (Faria-Oliveira et al., 1978), flavins (Haun et al., 1978), biacetyl (Bechara et al., 1979), xanthene dyes (Duran & Cilento, 1980), and chlorophyll-*a* (Brunetti et al., 1983); Nassi & Cilento, 1983). The latter process, promoting of photochemical reactions, was observed as phytochrome phototransformations (Augusto et al., 1978), chlorpromazine photooxidation (Duran et al., 1978), single-strand breaks in DNA (Meneghini et al., 1978), tryptophan photooxidation (Rivas-Suarez & Cilento, 1981), conversion of colchicine into lumocolchicines (Brunetti et al., 1982), and promotion of binding of riboflavin to lysozyme (Duran et al., 1983).

The aerobic oxidation of other substrates catalyzed by horseradish peroxidase was also shown to produce triplet carbonyls, which can accomplish photochemical-like processes (Cilento, 1984). Another aspect is provided by the peroxidase-catalyzed oxidation of indole-3-acetic acid which, in addition to triplet carbonyl compound, yields $^1O_2$ by quenching of the latter by ground state molecular O_2 (Vidigal et al., 1979).

The formation of an intermediate dioxetane was also proposed to explain the chemiluminescence arising from the metabolism of benzo(a)pyrene by liver microsomal fractions. This photoemission is produced concomitantly with the oxygenation to the ultimate carcinogenic metabolite benzo(a)pyrene-7,8-dihydrodiol-9,10-epoxide. However, the generation of excited states from benzo(a)pyrene-7,8-diol was proposed to originate from a minor non-enzymatic pathway (perhaps $^1O_2$-mediated) resulting in the formation of an intermediate dioxetane which decomposes to an excited state aldehyde (Seliger et al., 1982). A cytochrome P-450 oxoferryl complex may mimic the $^1O_2$-mediated oxidation of benzo(a)pyrene-7,8-diol in the cellular milieu. A sensitive chemiluminescence assay for benzo(a)pyrene-7,8-diol was also proposed, based on the fact that the yield of the dioxetane intermediate can be enhanced 10^4-fold when luminescence is coupled to a $^1O_2$-generating system (Thompson et al., 1983).

Enzymatic Generation of Singlet Oxygen

Monomol emission of $^1O_2$ was observed during the lactoperoxidase-, chloroperoxidase-, and myeloperoxidase/H_2O_2/halide systems (Khan, 1983, 1984a,b; Kanofsky, 1983, 1984a,b). The mechanism to produce $^1O_2$ stoichiometrically from the lactoperoxidase system is known to proceed in a two-step reaction as a disproportionation of H_2O_2; the requirement of halide serves as a trigger of H_2O_2 decomposition with intermediate formation of hypohalous acid; a second molecule of H_2O_2 is decomposed by the latter with generation of free $^1O_2$ (reaction 11).

$$H_2O_2 + H^+ + Br^- \rightarrow H_2O + BrOH \tag{11a}$$

$$H_2O_2 + BrOH \rightarrow H_2O + H^+ + Br^- + {}^1O_2 \tag{11b}$$

The production of $^1O_2$ by these systems, therefore, is clearly similar to that of the classic inorganic H_2O_2/OCL^- reaction.

Another example is the prostaglandin-hydroperoxidase activity of prostaglandin-endoperoxide-synthase. The strong emission intensity arising from the arachidonate-supplemented ram seminal microsomes or purified PG-endoperoxide synthase (Marnett,

1974; Duran & Suwa, 1981) permitted the spectral study of the signal (Cadenas et al., 1983d). Two peaks at 634 and 703 nm were observed, with a minimal intensity at 668 nm; this, along with enhancement and quenching of emission intensity by DABCO and β-carotene, respectively, provides a reasonable indication for $^1O_2$ participation. The mechanism for $^1O_2$ generation during this reaction bears an analogy to the production of $^1O_2$ during the cytochrome P-450-catalyzed heterolytic cleavage of hydroperoxide (Cadenas et al., 1983c). The hypothetical mechanism for prostaglandin-hydroperoxidase (Fe^{3+})-catalyzed $^1O_2$ production — involving an oxoferril complex intermediate $((FeO)^{3+})$ — is summarized in reaction 12.

$$PGG_2 + Fe^{3+} \rightarrow PGH_2 + (FeO)^{3+} \tag{12a}$$

$$PPG_2 + (FeO)^{3+} \rightarrow PGH_2 + Fe^{3+} + {}^1O_2 \tag{12b}$$

It is noteworthy that reactions 11, 12, and the heterolytic cleavage of hydroperoxides by cytochrome P-450 proceed as dismutation reactions yielding $^1O_2$ as depicted in the reaction 13.

$$ROOH + ROOH \rightarrow ROH + ROH + {}^1O_2 \tag{13}$$

$^1O_2$ generation was also reported during the horseradish peroxidase-catalyzed oxidation of malondialdehyde, where $^1O_2$ seems to be generated outside the enzyme, in the bulk solution (Duran et al., 1984). $^1O_2$ might originate from the decomposition of a hydroperoxide via a Russell's mechanism (Fig. 6A). This system mimics the photosensitized action of methylene blue and furocoumarines on ribosomes (Singh & Ewing, 1978; Singh & Vadasz, 1978), which is known to proceed through $^1O_2$ (De Toledo et al., 1983).

Generation of excited states by other enzymatic reactions as xanthine oxidase, lipoxygenase, and microsomal peroxidase activity were discussed by Krinsky (1979) and Duran (1982). However, the formation of excited states during those enzymatic reactions could be regarded as representing a minor-, non-enzymatic pathway involving recombination of O_2 radical products or oxidation of substrates by these radicals.

CONCLUSIONS

Two aspects should be considered to analyze the relevance of the occurrence of excited states in biological process.

The one aspect regards the generation of excited states as a consequence of cellular oxidative damage. Thus, low-level chemiluminescence arising from oxidative conditions, for example during lipid peroxidation, only reflects the formation of excited species in the terminal steps of membrane damage. In this regard, the low-level chemiluminescence technique could be regarded, along with other methods to evaluate peroxidative conditions (malondialdehyde accumulation, diene conjugate, or fluorescent product formation), as a technique for the evaluation of membrane damage. Moreover, low-level chemiluminescence offers the advantage of continuous monitoring and high sensitivity and, in some cases, it could be used as a non-invasive method, similarly to detection of volatile hydrocarbons and glutathione release from cells.

The other aspect is concerned with a functional role of the excited states in biological systems to support photochemical reactions in the dark or trigger amplification mechanisms (Cilento, 1982, White et al., 1974); this, of course, provides a different situation to that of photochemical transformations promoted by a free sensitizer excited by light. There is ample evidence that enzyme-generated triplet carbonyl species can promote photochemical-like effects, including reaction with and/or energy transfer to macromolecules with informational value as well as selective excitation of macromolecule residues. The excited carbonyl-driven conversion of santonin to lumisantonin, among other possibilities examined by White et al. (1974), indicate that excited states could play a role in biological reactions. With regard to $^1O_2$ it was proposed that this species might play a role in the control of the phototactic reaction sign of *Anaeba variabilis* (Schuchart & Nultsch, 1984).

Acknowledgments. Supported by Deutsche Forschungsgemeinschaft Schwerpunkt "Mechanismen Toxischer Wirkungen von Fremdstoffen".

Abbreviations:

GSH, reduced glutathione
DBS, 9,10-dibromoanthracene-2-sulfonate
DBA, dibromoanthracene
DPA, diphenylanthracene

REFERENCES

Allen, R.C., 1980, Free radical production by reticuloendothelial cells, in "The Reticuloendothelial System," 2:309, A.J. Sbarra and R.R. Strauss, eds., Plennum Publishing Corp., Oxford.

Allen, R.C., 1982, Biochemiexcitation: chemiluminescence and the study of biological oxygenation reactions, in "Chemical and Biological Generation of Excited States," p. 309, W. Adam and G. Cilento, eds., Academic Press, New York.

Allen, R.C., and Lieberman, M.M., 1984, Kinetic analysis of microbe opsonification based on stimulated polymorphonuclear leukocyte oxygenation activity, *Infec. Immun.*, 45:475.

Augusto, O., Cilento, G., Jung, J., and Song, P.-S., 1978, Phototransformation of phytochrome in the dark, *Biochem. Biophys. Res. Commun.*, 83:963.

Aust, S.D., and Svingen, B.A., 1982, The role of iron in enzymatic lipid peroxidation, in "Free Radicals in Biology," 5:1, W.A. Pryor, ed., Academic Press, New York.

Barenboim, G.M., Domanskii, A.N., and Turoverov, K.K., 1969, "Luminescence of Biopolymers and Cells," Plenum Press, New York.

Bartoli, G.M., Müller, A., Cadenas, E., and Sies, H., 1983, Antioxidant efficiency of diethyldithiocarbamate on microsomal lipid peroxidation assessed by low-level chemiluminescence and alkane production, *FEBS Lett.*, 164:371.

Bechara, E.J.H., Faria Oliveira, O.M.M., Duran, N., Casadei de Baptista, R., and Cilento, G., 1979, Peroxidase-catalyzed generation of triplet acetone, *Photochem. Photobiol.*, 30:101.

Bodaness, R.S., 1982, The potential role of NADPH and cytoplasmic NADP-linked dehyrogenases in protection against singlet oxygen mediated cellular toxicity, *Biochem. Biophys. Res. Commun.*, 108:1709.

Bodaness, R.S., and Chan, P.C., 1977, Singlet oxygen as a mediator in the hematoporphyrin-catalyzed photooxidation of NADPH to NADP in deuterium oxide, *J. Biol. Chem.*, 252:8554.

Bodaness, R.S., and Chan, P.C., 1979, Ascorbic acid as a scavenger of singlet oxygen, *FEBS Lett.*, 105:195.

Boh, E.E., Baricos, W.H., Bernofsky, C., and Steele, R.H., 1982, Mitochondrial chemiluminescence elicited by acetaldehyde, *J. Bioenerg. Biomembr.*, 14:115.

Borg, D.C., and Schaich, K.M., 1984, Cytotoxicity from coupled redox cycling of autooxidizing xenobiotics and metals, *Israel J. Chem.*, 24:38.

Boveris, A., and Cadenas, E., 1982, Production of superoxide radicals and hydrogen peroxide in mitochondria, in "Superoxide Dismutase," 2:15, L.W. Oberley, ed., CRC Press, Boca Raton.

Boveris, A., Cadenas, E., and Chance, B., 1980, Low-level chemiluminescence of the lipoxygenase reaction, *Photobiochem. Photobiophys.*, 1:175.

Boveris, A., Cadenas, E., and Chance, B., 1981, Ultraweak chemiluminescence: a sensitive assay for oxidative radical reactions, *Fed. Proc. Fed. Am. Soc. Exp. Biol.*, 40:195.

Boveris, A., Fraga, C.G., Varsavsky, A.I., and Koch, O.R., 1983, Increased chemiluminescence and superoxide production in the liver of chronically ethanol-treated rats, *Arch. Biochem. Biophys.*, 227:534.

Bowman, D.F., Gillan, T., and Ingold, K.U., 1971, Kinetic applications of EPR spectroscopy. III. Self reactions of dialkyl nitroxide radicals, *J. Am. Chem. Soc.*, 93:6555.

Brunetti, I.L., Bechara, E.J.H., Cilento, G., and White, E.H., 1982, Possible *in vivo* formation of lumicolchicines from colchicine by endogenously generated triplet species, *Photochem. Photobiol.*, 36:245.

Brunetti, I.L., Cilento, G., and Nassi, L., 1983, Energy transfer from enzymically-generated triplet species to acceptors in micelles, *Photo. Chem. Photobiol.*, 38:511.

Burton, G.W., Joyce, A., and Ingold, K.U., 1983, Is vitamin E the only lipid soluble, chain-breaking antioxidant in human blood plasma and erythrocyte membranes?, *Arch. Biochem. Biophys.*, 221:281.

Cadenas, E., 1984, Biological chemiluminescence, *Photochem. Photobiol.*, 40:823.

Cadenas, E., 1985, Oxidative stress and formation of excited species, in "Oxidative Stress", p. 311, H. Sies, ed., Academic Press, London.

Cadenas, E., Boveris, A., and Chance, B., 1980a, Chemiluminescence of lipid vesicles supplemented with cytochrome *c* and hydroperoxide, *Biochem. J.*, 188:577.

Cadenas, E., Boveris, A., and Chance, B., 1980b, Low-level chemiluminescence of hydroperoxide-supplemented cytochrome *c*, *Biochem. J.*, 187:131.

Cadenas, E., Boveris, A., and Chance, B., 1984a, Low-level chemiluminescence of biological systems, in "Free Radicals in Biology," 6:221, W.A. Pryor, ed., Academic Press, San Diego.

Cadenas, E., Brigelius, R., and Sies, H., 1983a, Paraquat-induced chemiluminescence of microsomal fractions, *Biochem. Pharmacol.*, 32:147.

Cadenas, E., Ginsberg, M., Rabe, U., and Sies, H., 1984b, Estimation of alpha-tocopherol antioxidant activity in microsomal lipid peroxidation as detected by low-level chemiluminescence, *Biochem. J.*, 223:755.

Cadenas, E., Müller, A., Brigelius, R., Esterbauer, H., and Sies, H., 1983b, Effects of 4-hydroxynonenal on isolated hepatocytes. Studies on chemiluminescence response, alkane production, and glutathione status, *Biochem. J.*, 214:479.

Cadenas, E., and Sies, H., 1982, Low-level chemiluminescence of liver microsomal fractions initiated by t-butyl hydroperoxide. Relation to microsomal hemoprotein, oxygen dependence, and lipid peroxidation, *Eur. J. Biochem.*, 124:349.

Cadenas, E., and Sies, H., 1984, Low-level chemiluminescence as an indicator of singlet molecular oxygen in biological systems, *Methods Enzymol.*, 105:221.

Cadenas, E., Sies, H., Campa, A., and Cilento, G., 1984c, Electronically excited states in microsomal membranes: use of chlorophyll-a as an indicator of triplet carbonyls, *Photochem. Photobiol.*, 40:661.

Cadenas, E., Sies, H., Graf, H., and Ullrich, V., 1983c, Oxene donors yield low-level chemiluminescence with microsomes and isolated cytochrome P-450, *Eur. J. Biochem.*, 130:117.

Cadenas, E., Sies, H., Nastainczyk, W., and Ullrich, V., 1983d, Singlet oxygen formation detected by low-level chemiluminescence during the enzymatic reduction of prostaglandin G_2 to H_2, *Hoppe-Seyler's Z. Physiol. Chem.*, 364:519.

Cadenas, E., Varsavsky, A.I., Boveris, A., and Chance, B., 1980c, Low-level chemiluminescence of cytochrome c-catalyzed decomposition of hydrogen peroxide, *FEBS Lett.*, 113:141.

Cadenas, E., Varsavsky, A.I., Boveris, A., and Chance, B., 1981a, Ultraweak chemiluminescence of brain and liver homogenates induced by oxygen organic hydroperoxide, *Biochem. J.*, 198:645.

Cadenas, E., Wefers, H., and Sies, H., 1981b, Low-level chemiluminescence of isolated hepatocytes, *Eur. J. Biochem.*, 119:531.

Catalani, L.H., and Bechara, E.J.H., 1984, Quenching of chemiexcited triplet acetone by biologically important compounds in aqueous medium, *Photochem. Photobiol.*, 39:823.

Chou, P.T., and Khan, A.U., 1983, L-ascorbic acid quenching of singlet delta molecular oxygen in aqueous media: generalized antioxidant property of vitamin C, *Biochem. Biophys. Res. Commun.*, 15:932.

Cilento, G., 1980, Generation and transfer of triplet energy in enzymatic systems, *Acc. Chem. Res.*, 13:225.

Cilento, G., 1982, Electronic excitation in dark biological processes, in "Chemical and Biological Generation of Excited States", W. Adam and G. Cilento, eds., p. 278, Academic Press, New York.

Cilento, G., 1984, Generation of electronically excited triplet species in biochemical systems, *Pure & Appl. Chem.*, 56:1179.

Dahlgren, C., and Briheim, G., 1985, Comparison between the luminol-dependent chemiluminescence of polymorphonuclear leukocytes and of the myeloperoxidase-HOOH system: influence of pH, cations, and protein, *Photochem. Photobiol.*, 41:605

De Toledo, S., Duran, N., and Singh, H., 1983, Inactivation of E. coli ribosomes by the bioenergized system malondialdehyde/horseradish peroxidase, *Photobiochem. Photobiophys.*, 5:237.

Doleiden, F.H., Fahrenholtz, S.R., Lamola, A.A., and Trozzolo, A.M., 1974, Reactivity of cholesterol and some fatty acids towards singlet oxygen, *Photochem. Photobiol.*, 20:519.

Duran, N., 1982, Singlet oxygen in biological processes, in "Chemical and Biological Generation of Excited States", W. Adam and G. Cilento, eds., p. 345, Academic Press, New York.

Duran, N., and Cilento, G., 1980, Long-range triplet exciton transfer from enzyme generated triplet acetone to xanthene dyes, *Photochem. Photobiol.*, 32:113.

Duran, N., Farias Furtado, S.T., Faljoni-Alario, A., Campa, A., Brunet, J.E., and Freer, J., 1984, Singlet oxygen generation from the peroxidase-catalyzed aerobic oxidation of an activated $-CH_2-$ substrate, *J. Photochem.*, 25:285.

Duran, N., Haun, M., De Toledo, S.M., Cilento, G., and Silva, E., 1983, Binding of riboflavin to lysozyme promoted by peroxidase-generated triplet acetone, *Photochem. Photobiol.*, 37:247.

Duran, N., Haun, M., Faljoni, A., and Cilento, G., 1978, Photochemical oxidation of chlorpromazine in the dark induced by enzymatically generated triplet carbonyl compounds, *Biochem. Biophys. Res. Commun.*, 81:785.

Duran, N., and Suwa, K., 1981, The generation of excited states during the action of prostaglandin endoperoxide synthase from rabbit renal medula, *Rev. Latinoamer. Quim.*, 12:13.

Encinas, M.V., Lissi, E.A., and Olea, A.F., 1985, Quenching of triplet benzophenone by vitamins E and C and by sulfur-containing aminoacids and peptides, *Photochem. Photobiol.*, 42:347.

Encinas, M.V., and Scaiano, J.C., 1981, Reaction of benzophenone triplets with allylic hydrogens. A laser flash photolysis study., *J. Am. Chem. Soc.*, 103:6393.

Esterbauer, H., 1982, Aldehydic products of lipid peroxidation, in "Free Radicals, Lipid Peroxidation, and Cancer", D.C.H. McBrien and T.F. Slater, eds., p. 101, Academic Press, New York.

Fahrenholtz, S.R., Doleiden, F.H., Trozzolo, A.M., and Lamola, A.A., 1974, On the quenching of singlet oxygen by alpha-tocopherol, *Photochem. Photobiol.*, 20:505.

Faria Oliveira, O.M.M., Haun, M., Duran, N., O'Brien, P.J., O'Brien, C.R., Bechara, E.J.H., and Cilento, G., 1978, Enzyme-generated electronically excited carbonyl compounds. Acetone phosphorescence during the peroxidase-catalyzed aerobic oxidation of isobutanal, *J. Biol. Chem.*, 253:4707.

Faulkner, L.R., and Glass, R.S., 1982, Electrochemiluminescence, in "Chemical and Biological Generation of Excited States" (W. Adam and G. Cilento, eds.), p. 191, Academic Press, New York.

Foote, C.S., 1976, Photosensitized oxidation and singlet oxygen: consequences in biological systems, in "Free Radicals in Biology", W.A. Pryor, ed., Vol. II, p. 85, Academic Press, New York.

Foote, C.S., Ching, T.-Y., and Geller, G.G., 1974, Chemistry of singlet oxygen. XVIII. Rates of reaction and quenching of alpha-tocopherol and singlet oxygen, *Photochem. Photobiol.*, 20:511.

Foote, C.S., and Denny, R.W., 1968, Chemistry of singlet oxygen. VI. Quenching by β-carotene, *J. Am. Chem. Soc.*, 90:6233. Foote, C.S., Denny, R.W., Weaver, L., Chang, Y., and Peters, J., 1970, Quenching of singlet oxygen, *Ann. New York Acad. Sci.*, 171:139.

Foote, C.S., Shook, F.C., and Akaberli, R.B., 1981, Chemistry of superoxide anion. 4. Singlet oxygen is not a major product of dismutation., *J. Am. Chem. Soc.*, 102:2503.

Förster, T., 1959, Transfer mechanisms of electronic excitation, *Dis. Faraday Soc.*, 27:7.

Ginsberg, M., and Cadenas, E., 1985, Electronically excited states during diaphorase-catalyzed benzoquinone reduction, *Photobiochem. Photobiophys.*, 9:223.

Gisler, G.C., Diaz, J., and Duran, N., 1982, Electronically excited species in the spontaneous chemiluminescence of urine and its uses in the detection of pathological conditions, *Physiol. Chem. Phys.*, 14:335.

Gisler, G.C., Diaz, J., and Duran, N., 1983, Observations on blood plasma chemiluminescence in normal subjects and cancer patients, *Arq. Biol. Tecnol.*, 26:345.

Grams, G.W., and Eskins, K., 1972, Dye-sensitized photooxidation of tocopherols. Correlation between singlet oxygen reactivity and vitamin E activity, *Biochemistry*, 11:606.

Haun, M., Duran, N., and Cilento, G., 1978, Energy transfer from enzymically generated triplet compounds to the fluorescent state of flavins, *Biochem. Biophys. Res. Commun.*, 81:779.

Heikkila, R.E., and Cabbat, F.S., 1978, Chemiluminescence from autoxidizing 6-hydroxydopamine. The involvement of active forms of oxygen, *Photochem. Photobiol.*, 28:667.

Howard, J.A., and Ingold, K.U., 1968, Rate constants for the self-reactions of n- and sec-butyl peroxy radicals and cyclohexylperoxy radicals. The deuterium isotope effect in the termination of secondary peroxy radicals, *J. Am. Chem. Soc.*, 90:1056.

Inbar, S., Linschitz, H., and Cohen, S.G., 1982, Quenching and radical formation in the reaction of photoexcited benzophenone with thiols and thioethers (sulfides). Nanosecond flash studies, *J. Am. Chem. Soc.*, 104:1679.

Kanofsky, J.R., 1983, Singlet oxygen production by lactoperoxidase. Evidence from 1270 nm chemiluminescence, *J. Biol. Chem.*, 258:5991.

Kanofsky, J.R., 1984a, Singlet oxygen production by lactoperoxidase: halide dependence and quantitation of yield, *J. Photochem.*, 25:285.

Kanofsky, J.R., 1984b, Singlet oxygen production by chloroperoxidase-hydrogen peroxide-halide systems, *J. Biol. Chem.*, 259:5596.

Kellog, R.E., Mechanism of chemiluminescence from peroxy radicals, 1969, *J. Am. Chem. Soc.*, 91:5433.

Kepka, A.G., and Grossweiner, L.I., 1973, Photodynamic inactivation of lysozime by eosin, *PHotochem. Photobiol.*, 18:49.

Khan, A.U., 1981, Direct spectral evidence of the generation of singlet molecular oxygen in the reaction of potassium superoxide with water, *J. Am. Chem. Soc.*, 103:6516.

Khan, A.U., 1983, Enzyme system generation of singlet molecular oxygen observed directly by 1.0-1.8μm luminescence spectroscopy, *J. Am. Chem. Soc.*, 105:7195.

Khan, A.U., 1984a, Discovery of enzyme generation of $^1\Delta g$ molecular oxygen: spectra of $(0,0)^1\Delta g \rightarrow {}^3\Sigma g^-$ IR emission, *J. Photochem.*, 25:327.

Khan, A.U., 1984b, Myeloperoxidase singlet oxygen generation detected by direct infrared emission, *Biochem. Biophys. Res. Commun.*, 122:668.

Khan, A.U., and Kasha, M., 1970, Chemiluminescence arising from simultaneous transitions in pairs of singlet oxygen molecules, *J. Am. Chem. Soc.*, 92:3293.

Koka, P., and Song, P.-S., 1978, Protection of chlorophyll-a by carotenoid from photodynamic decomposition, *Photochem. Photobiol.*, 28:509.

Koppenol, W.H., 1976, Reactions involving singlet oxygen and superoxide anion, *Nature* (London), 262:420.

Kraljic, I., and Sharpatyi, V.A., 1978, Determination of singlet oxygen rate constants in aqueous solutions, *Photochem. Photobiol.*, 28:583.

Krasnovsky, A.A., Jr., Kagan, V.E., and Minin, A.A., 1983, Quenching of singlet oxygen luminescence by fatty acids and lipids. Contribution of physical and chemical mechanisms, *FEBS Lett.*, 155:233.

Krinsky, N.I., 1979, Biological roles of singlet oxygen, in "Singlet Oxygen", H.H. Wasserman and W.A. Murray, eds., p. 597, Academic Press, New York.

Krinsky, N.I., 1982, Photobiology of carotenoid protection, in "The Science of Photomedicine", J.D. Regan and J.A. Parrish, eds., p. 397, Plenum Press, New York.

Marnett, L.J., Slodawer, P., and Samuelsson, B., 1974, Light emission during the action of prostaglandin synthase, *Biochem. Biophys. Res. Commun.*, 60:1286.

Matheson, I.B.C., 1979, The absolute value of the reaction rate constant of bilirrubin with singlet oxygen in D_2O, *Photochem. Photobiol.*, 29:875.

Matheson, I.B.C., Etheridge, R.D., Kratowich, N.R., and Lee, J., 1975, The quenching of singlet oxygen by amino acids and proteins, *Photochem. Photobiol.*, 21:165.

Matheson, I.B.C., and Lee, J., 1979, Chemical reaction rates of aminoacids with singlet oxygen, *Photochem. Photobiol.*, 29:879.

McCarthy, M.-B., and White, R.E., 1983, Functional differences between peroxidase compound I and the cytochrome P-450 reactive oxygen intermediate, *J. Biol. Chem.*, 258:9153.

Meneghini, R., Hoffman, M.E., Duran, N., Faljoni, A., and Cilento, G., 1978, DNA damage during the peroxidase-catalyzed aerobic oxidation of isobutanal, *Biochem. Biophys. Acta*, 18:177.

Merenyi, G., Lind, J., and Eriksen, T.E., 1985, The reactivity of superoxide $(O_2^-\cdot)$ and its ability to induce chemiluminescence with luminol, *Photobiol. Photochem.*, 41:203.

Miyazawa, T., Kaneda, T., Takyu, C., and Inaba, H., 1983a, Characteristics of tissue ultraweak chemiluminescence in rats fed with autoxidized linseed oil, *J. Nutr. Sci. Vitaminol.*, 29:53.

Miyazawa, T., Kaneda, T., Takyu, C., Yamagishi, A., and Inaba, H., 1981, Generation of singlet molecular oxygen in rat liver homogenate on adding autoxidized linseed oil, *Agric. Biol. Chem.*, 45:1597.

Miyazawa, T., Nagaoka, A., and Kaneda, T., 1983b, Tissue lipid peroxidation and ultraweak chemiluminescence in rats dosed with methyl linoleate hydroperoxide, *Agric. Biol. Chem.*, 47:1333.

Miyazawa, T., Sato, C., and Kaneda, T., 1983c, Antioxidative effects of alpha-tocopherol and riboflavin-butyrate in rats dosed with methyl linoleate hydroperoxide, *Agric. Biol. Chem.*, 47:1577.

Miyazawa, T., Tsuchiya, K., and Kaneda, 1984, Riboflavin tetrabutyrate: an antioxidative synergist of alpha-tocopherol as estimated by hepatic chemiluminescence, *Nutr. Rep. Internat.*, 29:157.

Nakano, M., and Noguchi, T., 1977, Mechanism of the generation of singlet oxygen in NADPH-dependent microsomal lipid peroxidation, in "Biochemical and Medical Aspects of Active Oxygen", O. Hayaishi and K. Asada, eds., p. 29, University of Tokyo Press, Tokyo.

Nassi, L., and Cilento, G., 1983, Red emission from chloroplasts elicited by enzyme-generated triplet acetone and triplet indole-3-aldehyde, *Photochem. Photobiol.*, 37:233.

Nilsson, R., Merkel, P.B., and Kearns, D.R., 1972, Unambiguous evidence for the participation of singlet oxygen in photodynamic oxidation of amino acids, *Photochem. Photobiol.*, 16:117.

Ozawa, T., and Hanaki, A., 1983, Reactions of superoxide ion with tocopherol and its model compounds: correlation between the physiological activities of tocopherol and the concentration of chromanoxyl-type radicals, *Biochem. Intern.*, 6:685.

Peters, G., and Rodgers, M.A.J., 1980, On the feasibility of electron transfer to singlet oxygen from mitochondrial components, *Biochem. Biophys. Res. Commun.*, 96:770.

Quickenden, T.I., Comarmond, M.J., and Tilbury, R.N., 1985, Ultraweak bioluminescence spectra of stationary phase growth of Saccharomyces cerevisiae and Schizosaccharomyces pombe, *Photochem. Photobiol.*, 41:611.

Rivas-Suarez, E., and Cilento, G., 1981, Quenching of enzyme-generated acetone phosphorescence by indole compounds: stereospecific effects of D- and L-tryptophan. Photochemical-like effects, *Biochemistry*, 20:7329.

Rooney, M.L., 1983, Ascorbic acid as a photooxidation inhibitor, *Photochem. Photobiol.*, 38:619.

Russell, G.A., 1957, Deuterium-isotope effects in the autoxidation of aralkyl hydrocarbons. Mechanism of the interaction of peroxy radicals, *J. Am. Chem. Soc.*, 79:3871.

Salin, M.L., Quince, K.L., and Hunter, D.J., 1985, Chemiluminescence from mechanically-injured soybean root tissue, *Photobiochem. Photobiophys.*, 9:271.

Schuchart, H., and Nultsch, W., 1984, Possible role of singlet molecular oxygen in the control of the phototactic reaction sign of Anabaena variabilis, *J. Photochem.*, 25:317.

Schulte-Herbrüggen, T., and Cadenas, E., 1985a, Electronically excited state generation during the lipoxygenase-catalyzed aerobic oxidation of arachidonate. The effect of reduced glutathione, *Photobiochem. Photobiophys.*, 10:35.

Schulte-Herbrüggen, T., and Cadenas, E., 1985b, Formation of electronically excited states during the lipoxygenase-catalyzed oxidation of fatty acids with different degree of unsaturation, in "Free Radicals in Liver Injury", M.U. Dianzani, G. Poli, T.F. Slater, and K.H. Cheeseman, eds., p. 91, IRL Press Ltd, Oxford.

Seliger, H.H., 1960, A photoelectric method for the measurement of spectra of light sources of rapidly varying intensities, *Anal. Biochem.*, 1:60.

Seliger, H.H., Thompson, A., Hamman, J.P., and Poser, G.H., 1982, Chemiluminescence of benzo(a)pyrene-7,8-diol, *Photochem. Photobiol.*, 36:359.

Shenck, G.O., Gollnick, K., and Neumüller, O.A., 1957, Zur photosensitilisierten Autoxydation der Steroide. Darstellung von Steroid-Hydroperoxiden mittels phototoxischer Photosensitilisatoren, *Ann. Chem.*, 603:46.

Singh, H., and Ewing, D.D., 1978, Methylene blue sensitized photoinactivation of E. Coli Ribosomes: effect on the RNA and protein components, *Photochem. Photobiol.*, 28:547.

Singh, H., and Vadasz, J.A., 1978, Singlet oxygen: a major reactive species in the furocoumarin photosensitized inactivation of E. coli ribosomes, *Photochem. Photobiol.*, 28:539.

Slawinska, D., 1978, Chemiluminescence and the formation of singlet oxygen in the oxidation of certain polyphenols and quinones, *Photochem. Photobiol.*, 28:453.

Slawinska, D., and Slawinski, J., 1973, Chemiluminescence in the oxidation reactions of some quinones and their polymers. I. Characteristics of reactions and luminescence, in "Chemiluminescence and Bioluminesence", M.C. Cormier, D.M. Hercules, and J. Lee, eds., p. 490, Plenum Press, New York.

Slawinska, D., and Slawinski, J., 1983, Biological chemiluminescence, *Photochem. Photobiol.*, 37:709.

Slawinski, J., Galezowiski, W., and Elbanowski, 1981, Chemiluminescence in the reaction of cytochrome *c* with hydrogen peroxide, *Biochim. Biophys. Acta*, 637:130.

Smith, G.J., 1978, Photooxidation of tryptophan sensitized by methylene blue, *J. Chem. Soc. Faraday Trans. 2*, 74:1350.

Smith, G.J., 1983, Reaction of retinol and retinal with singlet oxygen, *Photochem. Photobiol.*, 38:119.

Snyakin, A.P., Samsonava, L.V., Shlyapinkokh, V.Ya., and Ershov, V.v., 1978, Kinetics and mechanism of the interaction of sterically-hindered phenol with singlet oxygen, *Bull. Acad. Sci. USSR Div. Chem. Sci.*, 27:46.

Stauff, J., and Bartolmes, P., 1970, Chemilumineszenz bei der oxidativen bildung von triplettzuständen des anthrasemichinon- und anthrachinon-2-sulfonats, *Angew. Chem.*, 82:321.

Stauff, J., Schmidkunz, H., and Hartmann, G., 1963, Weak chemiluminescence of oxidation reactions, *Nature*, 198:281.

Stevens, B., 1973, Kinetics of photoperoxidation in solution, *Acc. Chem. Res.*, 6:90.

Thompson, A., Seliger, H.H., and Posner, G.H., 1983, A chemiluminescence assay specific for the microsomal metabolite benzo(a)pyrene-7,8-dihydrodiol, *Anal. Biochem.*, 130:498.

Trush, M.A., Mimnaugh, E.G., Siddik, Z.H., and Gram, T.E., 1963, Bleomycinmetal interaction: ferrous iron-initiated chemiluminescence, *Biochem. Biophys. Res. Commun.*, 112:378.

Ullrich, V., 1977, in The mechanism of cytochrome P-450 action, "Microsomes and Drug Oxidations", V. Ullrich, I. Roots, A. Hildebrandt, R.O. Estabrook, and A.H. Conney, eds., p. 192, Plenum Press, New York.

Villablanca, M., and Cilento, G., 1985, Enzymatic generation of electronic excited states by electron transfer, *Photochem. Photobiol.*, 42:591.

Vladimirov, Yu.A., Olenev, V.I., Suslova, T.B., and Cheremesina, Z.P., 1980, Lipid peroxidation in mitochondrial membrane, *Adv. Lip. Res.*, 17:173.

Vidigal, C.C.C., Faljoni-Alario, A., Duran, N., Zinner, K., Shimizu, Y., and Cilento, G., 1979, Electronically excited species in the peroxidase-catalyzed oxidation of indole-3-acetic acid: effect upon DNA and RNA, *Photochem. Photobiol.*, 30:195.

Wagner, P.J., 1971, Type II photoelimination and photocyclization of ketones, *Acc. Chem. Res.*, 4:168.

Wefers, H., and Sies, H., 1983, Hepatic low-level chemiluminescence during redox cycling of menadione and the menadione-glutathione conjugate. Relation to glutathione and
NAD(P)H:quinone reductase (DT-diaphorase) activity, *Arch. Biochem. Biophys.*, 224:568.

Weil, 1965, On the mechanism of the photooxidation of aminoacids sensitized by methylene blue,
Arch. Biochem. Biophys., 110:57.

White, E.H., Miano, J.D., Watkins, C.J., and Breaux, E.J., 1974, Chemical excited states, *Angew.
Chem.*, 13:229.

White, E.H., and Wei, C.C., 1970, A possible role for chemically-produced excited states in biology, *Biochem. Biophys. Res. Commun.*, 39:1219.

Wu, K.C., and Trozzolo, A.M., 1979, Evidence for the production of singlet molecular oxygen
from the quenching of excited states of dialkyl ketones by molecular oxygen, *J. Photochem.*,
10:407.

THE SUPEROXIDE-GENERATING NADPH OXIDASE
OF WHITE BLOOD CELLS

Owen T.G. Jones

Department of Biochemistry
University of Bristol
Bristol BS8 1TD, U.K.

INTRODUCTION

Blood contains a number of cell types with very varied functions. Red blood cells (erythrocytes), which have the principal function of transferring oxygen to the tissues and carbon dioxide from the tissues to the lungs, are much the most abundant. The white blood cells (or leucocytes) make up about 0.1% of the cells of blood and are largely involved in the processes of identifying and killing infecting microorganisms (Table 1). As indicated in Table 1, one group of leucocytes, the monocytes, are capable of leaving the blood and develop further into a group of cells known as macrophages. They are found in such tissues as lung, peritoneum, liver (the Kupfer cells) where they can phagocytose cell debris and other particles and, perhaps, participate in the recognition and killing of foreign cells. Macrophages have different physical form and functions in different tissues. The eosinophils, neutrophils and monocytes and those macrophages concerned in direct microbicidal activity, release $O_2^-\cdot$ when stimulated, and this $O_2^-\cdot$ is important in microbicidal (Babior, 1984) and tumouricidal activity (Hafeman & Lucas, 1979).

Neutrophils are the most easily prepared phagocytes and much of this article will be devoted to describing their properties. They are also called polymorphonuclear leucocytes, because the very large multilobed nucleus is a dominating feature of electron micrographs of the cells, and are also known as granulocytes, because their cytoplasm is packed with membrane-limited granules. The granules contain a range of proteins which are likely to participate in the killing of microbes. Two types of granules can be readily separated by centrifugation of neutrophil homogenates on gradients of sucrose. The primary granules (also called azurophil granules) are particularly rich in myeloperoxidase; so much of this heme-protein is present in neutrophils that it gives them their characteristic slightly green colour. Other proteins which are found in the primary granules include lysozyme, which digests bacterial cell walls, a variety of lysosomal hydrolytic enzymes and a group of cationic proteins which are directly microbicidal. The specific (or

Table 1. Leucocytes found in human blood.

Cell Type	Relative Abundance (%)	Main Function	O_2^- Production
Neutrophils	65	Kill bacteria	Yes
Lymphocytes	28	Immune responses	No
Monocytes	6	Become macrophages	Yes
Eosinophils	2	Kill parasites	Yes
Basophils	0.5	Histamine release	No

These populations of leucocytes are not fixed and may change in infections. Platelets, which are very abundant in blood, have not been included.

secondary) granules also contain lysozyme and are peculiarly rich in lactoferrin, the iron-binding protein. The abundance of lactoferrin may serve to modulate iron-catalysed radical reactions but its role is yet to be satisfactorily explained: within the granules it is very largely free from bound iron.

Neutrophils, and other granulocytes, move out of the blood vessels into the tissues by passing through the junctions of the endothelial cells which line them. They move to sites of infection, attracted up gradients of molecules for which there are specific receptors on the plasma membrane. For neutrophils these chemoattractants include bacterial peptides (for which the peptide N-formyl-methionyl-leucylphenylalanine, FMLP, is a useful synthetic analogue), leukotrienes derived from arachidonic acid, the complement products C3a and C5a and lymphokines, proteins secreted by some leucocytes at the infection site.

Foreign cells, such as infecting bacteria, become coated with antibodies to their surface antigens and with complement fragments (C5a). When these bind to specific receptors on the neutrophil surface, a complex sequence of events is initiated. The microbe is engulfed by outgrowths of the neutrophil plasma membrane, leading to the formation of an internalised phagocytic vacuole containing the microbe. Granules move through the cytosol to fuse with the phagocytic vacuole membrane and empty their contents into the vacuole. Within a few minutes, the microbe is killed and its destruction is started. During the time of phagocytosis there is a sudden increase in the oxygen consumption of the phagocyte (Fig. 1) which is unrelated to mitochondrial respiration; it is insensitive to cyanide or antimycin A. The cytosol of neutrophils contains a very small population of mitochondria and the true respiratory rate of the phagocyte is low. The oxygen consumed in the "respiratory burst" is converted to a series of derivatives, many of which are extremely reactive and likely to contribute to the microbicidal activity of the neutrophil.

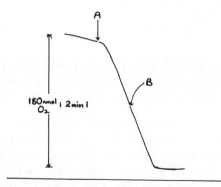

Figure 1. The respiratory burst of human neutrophils. A suspension of human neutrophils (10^8 cells ml^{-1}) at 37 °C in a Clark-type oxygen electrode was treated at A with 10^{10} latex particles opsonized with human immunoglobulin G. At B, KCN (1 mM final concentration) was added. The initial rate of oxygen uptake was 0.65 nmol $O_2/10^7$ cells; the stimulated rate was 10.0 nmol $O_2/10^7$ cells.

Products of the Respiratory Burst

The first product of the increased respiratory activity of the neutrophils is superoxide ($O_2^-\cdot$) (Babior, 1984). This one electron reduction of oxygen initiates the complex series of interactions shown below:

$$O_2 + e^- \rightarrow O_2^-\cdot \qquad E_o = -330mV \tag{1}$$

$$O_2^-\cdot + O_2^-\cdot + 2H^+ \rightarrow H_2O_2 + O_2 \tag{2}$$

$$O_2^-\cdot + H_2O_2 \rightarrow OH^- + OH\cdot + O_2 \tag{3}$$

Because reaction 1 has such a low mid point potential, the electron donor to the oxygen molecule must also have a low oxidation-reduction potential if superoxide is to be formed with good kinetics. The product, $O_2^-\cdot$, has a pK near 4.9 and so is largely charged at pH 7.0. Reaction 2 is catalysed by the enzyme superoxide dismutase which is abundant in the cytosol of mammalian cells, or by other metal catalysts such as Mn^{2+} or Cu^{2+}. The product, H_2O_2, has a pK around 11.0 and so is nearly fully protonated at physiological pH. Reaction 3, the Haber-Weiss reaction, proceeds poorly at physiological pH, and probably the production of OH· is catalysed by metal ions such as Fe^{3+} or Cu^{2+} in a Fenton reaction:

$$O_2^-\cdot + Fe^{3+} \rightarrow Fe^{2+} + O_2 \tag{4}$$

$$Fe^{2+} + H_2O_2 \rightarrow OH^- + OH\cdot + Fe^{3+} \tag{5}$$

As discussed elsewhere in this volume, the hydroxyl radical is highly reactive and will attack organic molecules immediately adjacent to its site of formation.

Further toxic products are derived from H_2O_2 by the action of myeloperoxidase released from the primary granule together with halogen ions:

145

$$H_2O_2 + Cl^- \xrightarrow{\text{myeloperoxidase}} OCl^- + H_2O \tag{6}$$

It was shown by Klebanoff (1967) that the hypochlorite formed in this reaction is several hundred-fold more microbicidal than H_2O_2 alone, and this molecule is likely to play a very important part in the neutrophils' attack on infecting microorganisms.

The formation of this group of very reactive derivatives of oxygen is initiated by the one electron reduction of oxygen to O_2^-. If this reaction is defective, then none of the reactive oxygen products can be formed. Such a defect is known to occur in human leucocytes, but is very rare and gives rise to the condition known as chronic granulomatous disease. This is a genetic disorder, with either X-linked or autosomal recessive inheritance. Neutrophils from patients with this disease fail to mount a respiratory burst and their capacity to kill microbes is seriously impaired. Such patients suffer from recurrent infections which, previous to the development of antibiotic therapy, led to early death. It is interesting that whilst failure to produce O_2^- leads to severe infections, the absence of myeloperoxidase from neutrophils, a genetic condition which is not uncommon, appears to cause few problems.

Protective Enzymes

Neutrophils are short-lived cells with a lifetime of perhaps less than 24h in the tissues, but contain a battery of enzymes in their cytosol which protect them from self-destruction through the action of activated oxygen. The sequence of protective reactions is listed below. O_2^- is first removed by dismutation:

$$O_2^- + O_2^- + 2H^+ \xrightarrow[\text{dismutase}]{\text{superoxide}} H_2O_2 + O_2 \tag{7}$$

This reaction catalysed by superoxide dismutase greatly reduces the concentration of O_2^- available as a substrate for any Fenton reaction. The H_2O_2 produced is broken down by the enzyme glutathione peroxidase, which requires glutathione (GSH) as substrate, yielding oxidised glutathione as product:

$$2GSH + H_2O_2 \xrightarrow[\text{peroxidase}]{\text{glutathione}} GSSG + 2H_2O \tag{8}$$

GSSG is re-reduced by the action of glutathione reductase, a cytosolic enzyme which requires NADPH as electron donor

$$GSSG + NADPH + H^+ \xrightarrow[\text{reductase}]{\text{glutathione}} 2GSH + NADP^+ \tag{9}$$

NADPH is in turn regenerated through the activity of the pentose phosphate pathway.

The heme protein catalase may also contribute to the removal of H_2O_2 by catalysing its decomposition to water and oxygen, but its relatively high K_m for H_2O_2 (around 1 mM) makes it a less effective protective enzyme than glutathione peroxidase. It is probable that vitamin E (α-tocophenol) and vitamin C (ascorbic acid) contribute some protection to the phagocyte against radical attack.

The life of a neutrophil at a scene of infection is very short and dead neutrophils are a major component of pus, together with their bacterial victims, but the anti-oxidant defences prolong its survival so that it can produce its quota of microbicidal missiles. There is evidence (Ohno et al., 1982) that radical production of the phagocyte is activated only in the region of plasma membrane surrounding the phagocytosed target, and so, if the population of target microbes is small, the antioxidant defences will readily prolong the survival of the phagocyte. This is particularly true of longer lived phagocytes such as macrophages which survive in the tissues for many months.

Properties of the Respiratory Burst Oxidase

A membrane fraction from the postnuclear supernatant of neutrophils which have been stimulated with immunoglobulin-coated particules or a variety of other stimuli catalyses the production of $O_2^-\cdot$ from oxygen if it is supplied with reduced pyridine nucleotide. The K_m for NADH (around 500 μM) is around ten times as great as the K_m for NADPH (about 40 μM) and it is likely that NADPH is the natural substrate (Iyer & Quastel, 1963). NADPH, which is the probable substrate of the respiratory burst oxidase and of glutathione reductase, which protects against its toxic products, is produced in neutrophils by the activity of enzymes of the pentose phosphate pathway. The reactions in this pathway which form NADPH are shown below:

$$\text{glucose-6-phosphate} + \text{NADP}^+ \xrightarrow[\text{dehydrogenase}]{\text{G-6-P}} \text{6-phosphoglucuronate} + \text{NADPH} \quad (10)$$

$$\text{6-phosphoglucuronate} + \text{NADP}^+ \xrightarrow[\text{dehydrogenase}]{\text{6-phosphoglucuronate}} CO_2 + \text{NADPH} +$$

$$\text{ribulose-5-phosphate} \quad (11)$$

The involvement of the pentose phosphate pathway in respiratory burst activity is demonstrated by the release of $^{14}CO_2$ from added 1-[^{14}C] glucose when neutrophils (or other phagocytes) are treated with a stimulus. The consumption of NADPH by activated phagocytes produces a raised concentration of NADP$^+$ which increases the activity of glucose-6-phosphate dehydrogenase and the whole pentose phosphate pathway.

The NADPH oxidase of leucocytes appears to be located in the plasma membrane fraction of the cells rather than in some intracellular organelle. The careful sucrose gradient fractionation of homogenates of activated cells showed that oxidase activity was associated with plasma membrane markers in neutrophils (Dewald et al., 1979) and in peritoneal macrophages (Sasuda et al., 1983; Berton et al, 1982). Confirmation of this localisation came from a different type of experiment involving the use of cytoplasts; these are plasma membrane vesicles empty of granules and of nuclei prepared by the centrifugation of neutrophils layered on a step gradient. Dense granule-packed vesicles separate from lighter granule-free vesicles at the interface of the carefully selected step gradient when centrifugal force is applied. The cytoplasts so prepared respond to stimuli and produce $O_2^-\cdot$ (Roos et al., 1983).

The plasma membrane location of the NADPH oxidase is an advantage in its function of producing $O_2^-\cdot$ in immediate proximity to a phagocytosed microbe. When phagocytosis is inhibited by treatment of neutrophils with cytochalasin B the yield of $O_2^-\cdot$ is increased, suggesting that the component of the oxidase which reacts with O_2 to form $O_2^-\cdot$ is on the outer face of the cytoplasmic membrane, and this becomes internalised during phagocytosis (see Figure 3 in later section).

Composition of NADPH-Oxidase

Neutrophils contain a cytochrome *b* with the remarkably low oxidation-reduction midpoint potential ($E_{m,7}$) of -245 mV (Cross et al., 1981). With such a low $E_{m,7}$ this cytochrome *b* could participate effectively in an electron transport system which functions in the production of $O_2^-\cdot$ ($E_{m,7}$ = -160 mV) and there is now considerable evidence for this function. This low potential cytochrome *b* is also present in eosinophils, macrophages and monocytes, all of which produce $O_2^-\cdot$, but is absent from the wide variety of mammalian cells which have been examined (Segal et al., 1981). In the X-linked form of chronic granulomatous disease, this cytochrome *b* is absent (Segal et al., 1983). In female carriers of the X-linked disease, the concentration of cytochrome *b* in the neutrophils is variable and is directly proportional to the capacity of the neutrophils to produce $O_2^-\cdot$ in response to a stimulus. When a suspension of whole cells is stimulated *anaerobically* the low potential cytochrome *b* becomes reduced: when air is stirred into these cells the cytochrome *b* is rapidly oxidised. This behaviour is expected if cytochrome *b* is a component of a complex which transfers electrons from NADPH to O_2. The capacity of the anaerobic cell suspensions to reduce the cytochrome when treated with a stimulus of the respiratory burst is absent in neutrophils of patients with the autosomal recessive variant of chronic granulomatous disease. In such neutrophils no $O_2^-\cdot$ is made in response to a stimulus, although the cytochrome *b* is present (Segal & Jones, 1980).

The cytochrome *b* has a dual location in neutrophils, where it is found in both the plasma membrane and in the specific granules (Segal & Jones, 1979; Borregaard et al., 1983). In macrophages the cytochrome is found exclusively in the plasma membrane (J.T. Hancock & O.T.G. Jones, unpublished work).

NADPH oxidase is believed to contain FAD as well as cytochrome *b*. Triton-solubilised preparations of the oxidase from human neutrophils form $O_2^-\cdot$ when supplied with NADPH, and their activity is stimulated by the addition of FAD (Gabig & Babior, 1979) which suggests that FAD has a role in the oxidase. Isolated plasma membranes from neutrophils contain non-covalently bound FAD (Cross et al., 1982; Light et al, 1981; Gabig, 1983) and the activity of the oxidase in post-nuclear supernatants of detergent extracts is inhibited by the FAD analogue 5-deaza FAD. The ratio of FAD to cytochrome *b* in isolated plasma membranes is close to 1:1, but in the membrane of isolated phagocytic vacuoles this ratio rises close to 2:1 (Cross, et al., 1982), indicating that the ratio of these components may not be fixed and that formation of the vacuolar membrane may involve selection of components of the plasma membrane.

A plausible arrangement of flavin and cytochrome in a oxidase complex is shown below. Microsomal NADPH-cytochrome P-450 reductase and NADH-cytochrome b_5 reductase are model complexes with similar arrangements of flavin and cytochrome.

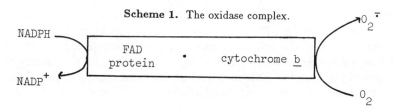

Scheme 1. The oxidase complex.

In the investigation of such a scheme, a supply of optically clear, very active solubilised NADPH oxidase is very important. This can be extracted from neutrophils prepared from fresh pig blood which have been previously activated with the soluble stimulus phorbol myristate acetate. Plasma membranes isolated from the stimulated cells are stirred gently with a mixture of 0.25% (v/v) Lubrol-PX and 0.25% (v/v) sodium deoxycholate which extracts over 50% of the NADPH oxidase with a specific activity of about 400 nmoles O_2^-.min^{-1}.mg protein^{-1} (Cross et al., 1984).

Analysis of the extracted oxidase shows that it contains almost equimolar amounts of FAD and cytochrome b; on addition of NADPH to an anaerobic solution of oxidase both FAD and cytochrome b become reduced, but the reduction is slow and incomplete, giving only about 50% reduction of the cytochrome after 30 minutes. Such slow rates of entry of electrons from NADPH into the anaerobic complex are much less than the rates necessary to make O_2^- when O_2 is present. It is likely that the presence of O_2 itself is necessary for the oxidase to receive electrons efficiently from NADPH. The evidence given below shows that in *aerobic* incubations the rate of reduction of cytochrome b and FAD by NADPH match precisely the measured rate of O_2^- production.

An anaerobic solution of NADPH-oxidase complex can be chemically reduced by titrating with the minimum of sodium ditionite solution, giving a reduced-minus-oxidised difference spectrum with a series of strong absorption bands at 559 nm, 530 nm and 427 nm characteristic of reduced cytochrome b (Figure 2), together with a trough at around 450 nm, due to the reduction of FAD. On mixing with O_2-saturated buffer in a stopped-flow spectrophotometer, the reduced cytochrome b of the dithionite-reduced complex is oxidised fast with a pseudo first order rate constant (k)=147s^{-1} (Cross et al., 1985). When NADPH was added to the oxidase preparation under aerobic conditions some cytochrome b was seen to be reduced: the steady state extent of reduction of the cytochrome was, typically, around 9% reduced (see Figure 2). In the steady state when electrons flow from NADPH to O_2 to make O_2^-, v, the calculated rate of cytochrome b oxidation (and of course, in a steady state this is equal to the rate of cytochrome b reduction) is given by the expression:

$$v = k \times \frac{[\text{cytochrome } b \text{ reduced}]}{[\text{total cytochrome } b]}$$

$$= 147s^{-1} \times \frac{9}{100}$$

$$= 13.23s^{-1}$$

For the same preparation the *measured* rate of O_2^- production on addition of NADPH was 13.03s^{-1} mol^{-1} cytochrome b, almost exactly the same as the *calculated* rate of electron flow into, and out of, cytochrome b. This indicates very strongly that every electron which flows from NADPH to O_2 passes through cytochrome b, and that the electron transport system of the oxidase is unbranched as shown in Scheme 1. This is an observation of some significance in determining the mechanism of the oxidase because it might be postulated that the FAD semiquinone, formed by one electron transfer from $FADH_2$ to cytochrome b, could react directly with O_2 to form $O_2^-\cdot$. Such a mechanism now seems very unlikely.

In agreement with the linear flow of electrons shown in Scheme 1, it was found that FAD was more reduced in the aerobic steady state than cytochrome b (40% reduced compared with 9%). The extent of reduction of both FAD and cytochrome b by NADPH in the steady state could be diminished by treatments which lowered the $O_2^-\cdot$-forming activity of the preparation: most conveniently by partial heat denaturation. Partly denatured oxidase demonstrated a linear relationship between extent of cytochrome b reduction and rate of $O_2^-\cdot$ production.

When NADPH is added to detergent extracts of plasma membranes from unstimulated neutrophils, no capacity for $O_2^-\cdot$ production is found. Similarly, no reduction of cytochrome b or FAD is observed in the aerobic steady state (Cross et al., 1985). This suggests that regulation of the oxidase acitivty *in vivo* is at, or before, the FAD component of the oxidase.

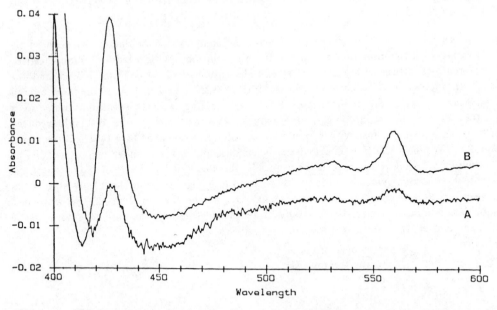

Figure 2. Difference spectra showing the effect of added NADPH on the steady state reduction of cytochrome b and FAD in solubilised NADPH oxidase from pig neutrophils. A baseline was obtained at 20°C with enzyme in both sample and reference cuvette. NADPH was added to the sample, and the spectrum A rapidly recorded in the aerobic steady state (expanded 4-fold here for clarity). The sample was fully reduced by addition of a few grains of sodium dithionite, and spectrum B was recorded.

Properties of the Oxidase and its Components

In neutrophils the oxidase complex forms about 5% of the total plasma membrane protein. The soluble oxidase is heat unstable with a half life of only 36s at 25 °C during turnover, and the cytochrome b appears to rapidly lose its heme in extracts at 30 °C. The cytochrome b binds CO, but with a low affinity (association constant, K = 1.18 mM, Cross et al., 1982) and also binds pyridine, isocyanides and imidazole, which act as inhibitors of the oxidase (Cross et al., 1984; Iizuka et al., 1985). The purified cytochrome b is a glycoprotein with M_r variously reported as 68,000-78,000 (Harper et al., 1984) 127,000 (Lutter et al., 1985) and 14,000, 12,000 and 11,000 (Pember et al., 1984). The E_m of the cytochrome b is pH dependent (Cross et al., 1981), which makes it a possible participant in proton translocation across the phagosome membrane, and is shifted upwards from -245 mV to -180 mV by treatment of the oxidase with the -SH reagent p-chloromercuribenzoate (PCMB). PCMB inhibits the NADPH oxidase activity and, at low concentrations, appears to affect the reduction of cytochrome b without affecting reduction of the FAD component.

The oxidase complex is inhibited by Mg^{2+}-binding chelating agents, but its function does not appear to require Ca^{2+}. It is inhibited by analogues of FAD, such as 5-deaza FAD and by quinacrine (50% inhibition at 1 mM). These inhibitors are relatively slow to act, and it appears that the FAD is not easily accessible for exchange with analogues. Supporting this view that the FAD is relatively protected is the finding that diaphorase activity of the complex is low, when ferricyanide or DCPIP are used as electron acceptors.

Other components of the oxidase complex have been proposed. In particular the involvement of ubiquinone has been suggested, particularly by Crawford & Schneider (1983) who found ubiquinone associated with the NADPH oxidase fraction in neutrophil homogenate fractionation. We have failed to confirm this report (Cross et al., 1983), and it has also been found that neutrophil cytoplasts, which apparently lack ubiquinone, are competent in the production of O_2^-· (Lutter et al., 1984). These conflicts should be capable of fairly early resolution when purified oxidase is more easily available.

The measured stoichiometry of the solubilised NADPH oxidase fits very closely to that expected from the equation below, when measurements of NADPH and O_2 consumption and O_2^-· production are made.

$$NADPH_2 + 2O_2 \rightarrow 2O_2^- · + NADP^+ \qquad (12)$$

There is no evidence for the direct formation of H_2O_2 by the oxidase. The K_m for NADPH is 45 μM, for NADH is 460 μM (Cross et al., 1984).

The oxidase has not been purified to homogeneity: indeed it may be difficult to do so without separating subunits of the complex and thus losing activity. There is some disagreement about the molecular weights of the most abundant peptides in the published purifications. In one purification (Markert et al., 1985) it was reported that in human NADPH oxidase the major component had a M_r = 65,000. A purified fraction (Bellevite et al., 1984; Papini et al., 1985) of pig or guinea pig neutrophil oxidase was greatly enriched in a peptide with a M_r = 31,000.

Although circulating phagocytes contain ample NADPH and are surrounded by O_2, they do not produce O_2^-· until they meet an appropriate stimulus. This stimulus transforms the oxidase from the resting state to its active form, usually after a lag period of some 30s. A surprising variety of particulate and soluble stimulants activate the respiratory burst of neutrophils, some of them being listed in Table 2. The first four are likely to be present at sites of infection. For the immune complexes, complement C5a and N-formyl peptides, there are specific receptors located on the plasma membrane, and binding of the stimulus to its appropriate receptor initiates activation of the oxidase. Using radiolabelled N-formyl peptides, affinity labelling of the FMLP receptor and its subsequent electrophoretic separation have been achieved (Neidel et al., 1980; Schmitt et al., 1983). The receptor appears to have two components of $M_r = 50,000$ and $M_r = 60,000$; and there are about 50,000 receptors per neutrophil.

Table 2. Some stimulants of the respiratory burst of neutrophils.

Immune complexes and antibody-coated particles

Complement fraction C5a

N-formyl peptides (e.g., FMLP)

Leukotriene B4

Endotoxin

Concanavalin A

Some detergents (e.g., digitonin)

Phorbol myristate acetate (PMA)

Arachidonic acid

NaF

Ca^{2+}-ionophores (e.g., A 23187)

Phagocytosable particles (e.g., latex beads, uric acid crystals)

Charged surfaces (e.g., glass)

A detailed discussion of the mechanism of activation of the oxidase by the wide variety of stimulants shown in Table 2 is beyond the scope of this review, but some general principles for receptor mediated activation are illustrated in Figure 3. The best characterised stimulus is the N-formyl peptide, FMLP, which on binding to its receptor initiates the receptor-mediated hydrolysis of the plasma membrane phospholipid phosphatidyl inositol *bis* phosphate (PIP_2) by activating a specific phosphodiesterase (a phospholipase C). Two products of the hydrolysis are signalling molecules: these are diacylglycerol (DAG) and inositol tris phosphate (IP_3). DAG is an activator of Ca^{2+}-dependent protein kinase C which may activate the oxidase by phosphorylation of one or more of its component polypeptides. IP_3 is believed to interact with intracellular receptors and

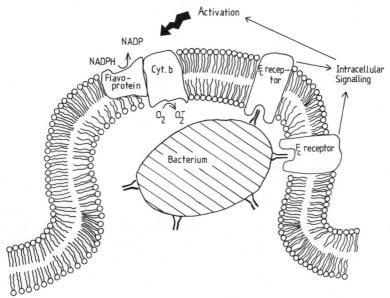

Figure 3. Initial events in the formation of a phagolysosome and activation of NADPH oxidase when the neutrophil plasma membrane encounters an IgG coated bacterium. The IgG binds to its specific receptor at the outer face of the cytoplasmic membrane, which leads to generation of a signal that activates the NADPH oxidase complex so that it can now accept electrons from NADPH. The O_2^-· is released extracellularly into the developing phagolysosome.

promote the release of Ca^{2+} from intracellular stores (Berridge & Irvine, 1984). This increased cytosolic Ca^{2+} is also an activator of protein kinase C.

Stimulants other than FMLP are likely to act by promoting various processes on this pathway. Phorbol myristate acetate is an analogue of DAG and activates protein kinase C, and causes the phosphorylation of membrane proteins without causing a rise in intracellular Ca^{2+}, and in the absence of extracellular Ca^{2+}. Phagocytosable stimuli also work without causing a significant rise in intracellular free Ca^{2+}.

It is possible that NaF activates by inhibiting the endogenous phosphatase which removes phosphate from activated oxidase. Ca^{2+} ionophores almost certainly act by permitting the entry of extracellular Ca^{2+} into the neutrophil to activate the phosphodiesterase and/or protein kinase C.

One observation that may be of clinical importance is that crystals of uric acid are potent activators of the respiratory burst and so are various crystal forms of calcium phosphate (Higson & Jones, 1985). Uric acid accumulates in joints in gout, which are a rich source of phagocytic cells, and activation of the neutrophils will cause the release of reactive oxygen derivatives and contribute to the inflammatory response and tissue damage. In a similar fashion, the immune complexes which are abundant in the blood in rheumatoid arthritis patients may activate phagocytes and so contribute to joint inflammation. These harmful effects of the NADPH oxidase may be moderated by therapeutic agents designed with an understanding of the mechanism of the oxidase and its activation.

REFERENCES

Babior, B.M., 1984, Oxidants from phagocytes: agents of defence and destruction, *Blood*, 64:959.

Bellavite, P., Jones, O.T.G., Cross. A.R., Papini, E., and Rossi, F., 1984, Composition of partially purified NADPH oxidase from pig neutrophils, *Biochem. J.*, 223:639.

Berridge, M.J., and Irvine, R.F., 1984, Inositol trisphosphate, a novel second messenger in cellular signal transduction, *Nature*, 312:315.

Berton, G., Bellavite, P., DeNicola, G., Dri, P., and Rossi, F., 1982, Plasma membrane and phagosome localisation of the activated NADPH oxidase in elicited peritoneal macrophages of the guinea pig., *J. Path.*, 136:241.

Crawford, D.R., and Schneider, D.L., 1983, Ubiquinone content and respiratory burst activity of latex filled phagolysosomes isolated from human neutrophils and evidence for the probable involvement of a third granule, *J. Biol. Chem.*, 258:5367.

Cross, A.R., Jones, O.T.G., Harper, A.M., and Segal, A.W., 1981, Oxidation reduction properties of the cytochrome *b* found in the plasma membrane of human neutrophils: a possible oxidase in the respiratory burst, *Biochem. J.*, 194:599.

Cross, A.R., Jones, O.T.G., Garcia, R., and Segal, A.W., 1982, The association of FAD with the cytochrome *b*-245 of human neutrophils, *Biochem. J.*, 208:759.

Cross, A.R., Jones, O.T.G., Garcia, R., and Segal, A.W., 1983, The subcellular localisation of ubiquinone in human neutrophils, *Biochem. J.*, 216:765.

Cross, A.R., Parkinson, J.F., and Jones, O.T.G., 1984, The superoxide-generating oxidase of leucocytes: NADPH-dependent reduction of flavin and cytochrome *b* in solubilised preparations, *Biochem J.*, 223:337.

Cross, A.R., Parkinson, J.F., and Jones, O.T.G., 1985, Mechanism of the superoxide-producing oxidase of neutrophils: O_2 is necessary for the fast reduction of cytochrome *b*-245 by NADPH, *Biochem. J.*, 226:881.

Dewald, B., Baggiolini, M., Curnutte, J.T., and Babior, B.M., 1979, Subcellular localisation of the superoxide-forming enzyme in human neutrophils, *J. Clin. Invest.*, 61:21.

Gabig, T., 1983, The NADPH-dependent O_2^--generating oxidase from human neutrophils. Identification of a flavoprotein component that is deficient in a patient with C.G.D., *J. Biol. Chem.*, 258:6352.

Gabig, T., and Babior, B.M., 1979, The O_2^--forming oxidase responsible for the respiratory burst in human neutrophils. Properties of the solubilised enzyme, *J. Biol. Chem.*, 254:9070.

Hafeman, D.E., and Lucas, Z.J., 1979, Polymorphonuclear leukocyte-mediated antibody-dependent cellular cytotoxicity against tumour cells. Dependence of oxygen and the respiratory burst, *J. Immunol.*, 123:55.

Harper, A.M., Dunne, M.J., and Segal, A.W., 1984, Purification of cytochrome *b*-245 from human neutrophils, *Biochem. J.*, 219:519.

Higson, F.K., and Jones, O.T.G., 1984, Oxygen radical production by horse and pig neutrophils induced by a range of crystals, *J. Rheumatol.*, 11:735.

Iizuka, T., Kanegasaki, S., Makino, R., Tanaka, T., and Ishimura, Y., 1985, Phyridine and imidazole reversibly inhibit the respiratory burst in porcine and human neutrophils: evidence for the involvement of cytochrome *b*-558 in the reaction, *Biochem. Biophys. Res. Commun.*, 130:621.

Iyer, G.Y.N., and Quastel, J.H., 1963, NADPH oxidase, *Can. J. Biochem. Physiol.*, 41:427.

Klebanoff, S.J., 1967, Iodination of bacteria: A bactericidal mechanism, *J. Exp. Med.*, 126:1063.

Light, D.R., Walsh, C., O'Callaghan, A.M., Goetzl, E.J., and Tauber A.I., 1981, Characteristics of the cofactor requirements of the O_2^--generating NADPH oxidase of human PMN leucocytes, *Biochemistry*, 20:1468.

Lutter, R., van Zweiten, R., Weening, R.S., Hamers, M.N., and Roos, D., 1984, Cytochrome *b*, flavins and ubiquinone-50 in enucleated human neutrophils (polymorphonuclear leucocyte cytoplasts), *J. Biol. Chem.*, 259:9603.

Markert, M., Glass, G.A., and Babior, B.M., 1985, Respiratory burst oxidase from human neutrophils: purification and some properties, *Proc. Natl. Acad. Sci.*, 82:3144.

Neidel, J.E., Davis, J., and Cuatrecasas, P., 1980, Covalent affinity labelling of the foryl peptide chemotactic receptor, *J. Biol. Chem.*, 255:7063.

Ohno, Y-I., Hirai, K-I., Kanoh, T., Uchino, H., and Ogawa, K., 1982, Subcellular localisation of H_2O_2 production in human neutrophils stimulated with particles and an effect of cytochalasin B on the cells, *Blood*, 60:253.

Papini, E., Grzeskowiak, M., Bellavite, P., and Rossi, F., 1985, Protein kinase C phosphorylates a component of NADPH oxidase of neutrophils, *FEBS Lett.*, 190:204.

Pember, S.O., Heyl, B.L., Kinkade, J.M., and Lambeth, J.D., 1984, Cytochrome *b*-558 from bovine granulocytes: partial purification from Triton X-114 extracts and properties of the isolated cytochrome, *J. Biol. Chem.*, 259:10590.

Roos, D., Voetman, A.A., and Meerhof, L.J., 1983, Functional activity of enucleated human polymorphonuclear leucocytes, *J. Cell Biol.*, 97:368.

Sasada, M., Pabst, M.J., and Johnston, R.B., 1983, Activation of mouse peritoneal macrophages by lipopolysaccharide alters the kinetic properties of the O_2^- producing NADPH-oxidase, *J. Biol. Chem.*, 258:9631.

Schmitt, M., Painter, R.G., Jesaitis, A.J., Preissner, K., Sklar, L.A., and Cochrane, C.G., 1983, Photoaffinity labelling of the N-formyl receptor binding site of intact human polymorphonuclear leukocytes. Evaluation of a label as suitable to follow the fate of the receptor ligand complex, *J. Biol. Chem.*, 258:649.

Segal, A.W., and Jones, O.T.G., 1979, The subcellular distribution and some properties of the cytochrome *b* component of the microbicidal oxidase system of human neutrophils, *Biochem. J.*, 182:181.

Segal, A.W., and Jones, O.T.G., 1980, Absence of cytochrome *b* reduction in stimulated neutrophis from both female and male patients with CGD, *FEBS Lett.*, 110:111.

Segal, A.W., Garcia, R., Goldstone, A.H., Cross, A.R., and Jones, O.T.G., 1981, Cytochrome *b*-245 of neutrophils is also present in human monocytes, macrophages and eosinophils, *Biochem. J.*, 196:363.

Segal, A.W., Cross, A.R., Garcia, R., Borregaard, N., Valerius, N.H., Soothill, J.F., and Jones, O.T.G., 1983, Absence of cytochrome *b*-245 in CGD. A multicentre European evaluation of its incidence and relevance, *New Eng. J. Med.*, 308:245.

GENERAL READING

"Haematology," by Williams, Beutter, Ersley, Rundles, McGraw Hill Book Co., New York, 1983; for general coverage of granulocyte, monocyte and macrophage functions.

"Molecular Biology of the Cell," by Alberts, Bray, Lewis, Raff, Roberts & Watson, Garland, New York & London, 1983. See sections on phagocytosis, differentiation and the immune system.

Babior, B., 1984, The respiratory burst of phagocytes, *J. Clin. Invest.*, 73:599. A very short review.

Berridge, M.J., 1984, Inositol triphosphate and diacylglycerol as second messengers, *Biochem. J.*, 220:345.

Bennett, J.P., Cockcroft., S., Caswell, A.H., and Gomperts, B.D., 1982, Plasma membrane location of phosphatidylinositol hydrolysis in rabbit neutrophils stimulated with formylmethionyl-leucylphenylalanine, *Biochem. J.*, 208:801.

Korchak, H.M., Vienne, K., Rutherford, L.E., and Weissman, G., 1984, Neutrophils stimulation: receptor, membrane and metabolic events, *Fed. Proc.*, 43:2743.

Cochrane, C.G., 1984, Mechanisms coupling stimulation and functions in leukocytes, *Fed. Proc.*, 43:2729.

Hitzig, W.H., and Seger, R.A., 1983, Chronic granulomatous disease, a heterogeneous syndrome, *Human Genetics*, 64:207.

ROLE OF REACTIVE OXYGEN SPECIES AND LIPID PEROXIDATION

IN CHEMICALLY INDUCED TOXICITY AND CARCINOGENESIS

Martyn T. Smith

Department of Biomedical and Environmental Health Sciences
School of Public Health
University of California
Berkeley, California 94720

INTRODUCTION

The multiple types of reactive oxygen species and their chemistry is reviewed elsewhere in this text. Superoxide anion radical ($O_2^-\cdot$), hydrogen peroxide (H_2O_2), hydroxyl radical ($HO\cdot$) and singlet oxygen ($^1O_2$) have been proposed as playing important roles in the toxicity of various chemical agents. Some examples are given in Table 1. Their role is, however, highly controversial and, dependent on your point of view, can be considered as being either of major or minor importance in toxicology. Here I propose to outline some of the evidence which indicates that reactive oxygen species play an important role in the toxic effects of various chemical compounds. For example, it is now apparent that reactive oxygen species can, (1) damage DNA, so as to cause mutation and chromosomal damage; (2) initiate the peroxidation of membrane lipids; (3) oxidize cellular thiols; (4) degrade polysaccharides; and (5) inhibit key enzymes; etc. Given these findings, it does not seem too difficult to imagine that chemicals which enhance the formation of reactive oxygen species would wreak havoc in living cells. One must always remember, however, that the cells of aerobic organisms are well protected against reactive oxygen species by a battery of enzymatic and non-enzymatic antioxidant defenses (for review cf. Sies, 1985) since the conversion of dioxygen to water in the respiratory process generates $O_2^-\cdot$, H_2O_2, and $HO\cdot$. Thus, reactive oxygen species are "normal" cellular metabolites in respiring cells and must be dealt with accordingly. A critical balance therefore exists between the generation and detoxication of reactive oxygen species in respiring cells. Environmental chemicals, such as drugs, pollutants, pesticides, herbicides, dietary components, etc., could disrupt this critical balance by one of two mechanisms: (a) inhibition or disruption of the cellular antioxidant defenses; and (b) marked stimulation of the formation of reactive oxygen species.

Here I propose to concentrate mainly on chemical agents which act in the latter fashion, but there are clear examples of the former. These include the cancer chemotherapeutic drug 1,3-bis(2-chloroethyl)-1-nitrosourea (BCNU) and aminotriazole, a commonly used herbicide. BCNU is a potent inhibitor of the enzyme glutathione reductase, a key component of the protective glutathione-dependent H_2O_2 detoxication system and aminotriazole is a potent inhibitor of catalase.

Table 1. Chemicals whose toxic effects may be mediated via the generation of reactive oxygen species.

Chemical	Target Organ, Effect
Alloxan	Pancreatic β - cell
Adriamycin	Heart, Antitumor agent
Divicine	Blood cells
6-Hydroxydopamine	Adrenergic nerve terminals
Menadione	Blood cells, liver
MPTP	Neurones of Subtantia Nigra
Nitrofurantoin	Lung
Paraquat	Lung
Streptonigrin	Antitumor agent
- Lapachone	Antimicrobial agent
Diethylstibestrol	Kidney, Uterus, Vagina (Cancer?)
Benzo (a) pyrene quinones	Lung, Stomach (Cancer?)
Gossypol	Sperm, Antitumor agent

REDOX CYCLING AND THE STIMULATION OF ACTIVE OXYGEN FORMATION IN CELLS

Most chemicals which stimulate the formation of reactive oxygen species in cells do so by a process known as redox cycling. This subject has been extensively reviewed (e.g., Kappus and Sies, 1981; Hochstein, 1983; Bus and Gibson, 1984; Smith et al., 1985). The process of redox cycling is illustrated in Figure 1. Briefly, the chemical agent (Q) is firstly reduced by one electron to its free radical form (Q·) and then oxidized by dioxygen back to the native compound (Q) with the concommitant formation of O_2^-·. This process can be non-enzymatic or may be catalyzed by numerous cellular flavoproteins, including NADPH-cytochrome P-450 reductase and NADH:ubiquinone oxidoreductase. Given a good supply of electrons to the flavoprotein and oxygenated conditions, a small amount of chemical can produce a large amount of O_2^-·, and subsequently H_2O_2, by redox cycling since none of the chemical is consumed in the process. If sufficient quantities of O_2^-· and H_2O_2 are produced so that the endogenous defense enzymes are overwhelmed, a "toxic active oxygen cascade" may be set in motion (Fig. 1). In this "cascade", reactive oxygen species cause cumulative damage to the cell in a process which is catalyzed by transition metals, such as Fe^{2+} and Cu^+ (Fig. 1). A certain "threshold of detoxication" must therefore be exceeded if this damage is to be sufficient to result in cell death and/or neoplasia.

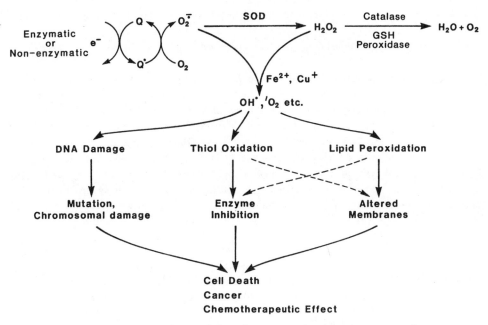

Figure 1. Redox cycling and the subsequent toxic active oxygen cascade.

Let us now consider what toxic insults may ensue once the "threshold of detoxication" has been crossed.

REACTIVE OXYGEN SPECIES ARE MUTAGENIC AND CLASTOGENIC

Exposure of both prokaryotic and eukaryotic cells to increased oxygen tension causes mutations in the DNA and chromosomal damage. Physiological oxygen concentrations have also been found to be mutagenic in certain oxygen-sensitive auxotrophs of *Salmonella typhymurium* strain TA 100 (Bruyninckx et al., 1978). Reactive oxygen species are almost certainly responsible for these DNA damaging effects of oxygen, since O_2^-· causes DNA strand scission and H_2O_2 liberates DNA bases, causes DNA strand breakage and alters the chemical structure of the bases. Weitzman, Stossel and co-workers have also shown that human phagocytes can produce mutations in bacteria in addition to mutation and sister chromatid exchanges in cultured hamster ovary cells and that reactive oxygen species are responsible for these effects (cf. Weitzman et al., 1985). Potassium superoxide also causes chromosome damage in chinese hamster ovary cells.

There is a strong relationship between mutagenicity and carcinogenicity (Ames and McCann, 1981). Two new Salmonella tester strains, TA102 and TA104, were recently introduced by Ames' group and they have described how various oxidants, including

H_2O_2 and several redox cycling quinones, are mutagenic to these bacteria (cf. Levin et al., 1982; Chesis et al., 1984). Various products of lipid peroxidation, including peroxides and aldehydes, are also mutagenic to these bacterial strains (Marnett et al., 1985).

REACTIVE OXYGEN SPECIES AND TUMOR PROMOTION

Inflammation as a result of irritation has long been known to play an important role in promoting cancer. Since reactive oxygen species are produced in copious amounts by phagocytes during the inflammatory process, it seems likely from the above discussion that these play a role in tumor promotion. The experimental evidence for this role has recently been reviewed in detail (Kensler and Trush, 1984; Troll and Weisner, 1985) and I will not discuss it in depth here. Essentially three main lines of evidence implicate a role for active oxygen in tumor promotion: (1) Many different classes of tumor promoters cause the generation of reactive oxygen species either directly or indirectly; (2) Tumor promoters lower cellular antioxidant defenses at least in the mouse skin model of carcinogenesis; and (3) Reactive oxygen scavengers and detoxifiers inhibit the effects of many tumor promoters *in vitro* and *in vivo*. It should be pointed out, however, that some workers consider that reactive oxygen species play little or no role in tumor promotion (Schwarz et al., 1984) and the two-stage model of skin carcinogenesis is itself highly controversial. Much of this controversy stems from the fact that the initiation-promotion sequence can be, at least, partially inverted (cf. Fürstenberger et al., 1985).

As discussed above, malignancies often develop at the sites of inflammation where phagocytic cells accumulate. In addition to the direct effects that the reactive oxygen species produced by these cells may have, there is also the possibility that they may be involved in the localized metabolism of chemicals to reactive intermediates (Trush et al., 1985). In this case the $O_2^-\cdot$ produced by phagocytes, after dismutation of H_2O_2, serves as a substrate for peroxidase enzymes which are also released from phagocytes during the oxidative burst. The reactive metabolites are formed when the parent compound acts as a reducing cofactor to reduce the peroxidase enzyme from its +5 to +3 state. This phenomenon could partly explain the relationship between inflammation and cancer as well as explain the site specificity of tumors caused by several classes of chemical carcinogens.

Clearly then, reactive oxygen species could play a major role in human carcinogenesis, acting at many stages in what is considered to be a multistage process. The critical macromolecular targets for the reactive oxygen remain undefined, as does the exact nature of the species damaging the cellular macromolecules. These are therefore important areas for future research.

ROLE OF REACTIVE OXYGEN SPECIES IN THE CYTOTOXICITY OF REDOX CYCLING COMPOUNDS

In addition to their chronic, DNA damaging effects in cells, reactive oxygen species may also be acutely toxic to cells if produced in sufficient quantities. One mechanism for their toxicity is the peroxidation of membrane lipids, but it must be emphasized that this is only *one* of many possible mechanisms. Others include the oxidation of protein thiols and the inactivation of key enzymes involved in maintaining normal cellular homeostasis (cf. Fig. 1). In Table 1, I have listed some examples of chemicals which may be toxic via the generation of reactive oxygen species. Using two examples, menadione and diquat, let us now explore how active oxygen may mediate the cytotoxic effects of these two compounds.

Menadione

The mechanism(s) of menadione (2-methyl-1,4-naphthoquinone) cytotoxicity has been extensively studied in our laboratory and in those of Orrenius and Sies. We recently reviewed these studies in some detail (Smith et al., 1985). Toxic concentrations of menadione cause the rapid depletion of cellular glutathione (GSH) and the formation of numerous small blebs on the surface of isolated cells (Thor et al., 1982). Many other toxic agents, as well as anoxia, cause membrane blebbing, which appears to be a common, early sign of toxic injury. Blebbing is almost certainly a result of disruption of the microfilament component of the cytoskeleton and its association with the plasma membrane (cf. Smith and Orrenius, 1984). This may come about by either a direct effect on the cytoskeletal components or by a disruption of intracellular ion homeostasis, most notably calcium ion homeostasis. Menadione has been shown to disrupt Ca^{2+} homeostasis within liver cells by inhibiting the activity of key Ca^{2+}-transporting enzymes in the endoplasmic reticulum and plasma membrane and by causing the release of mitochondrial Ca^{2+} stores. The oxidation of key protein thiols is clearly involved in these effects and is thought to be mediated largely by the production of reactive oxygen species, although direct arylation of cellular thiols by menadione also occurs (DiMonte et al., 1984); Orrenius et al., 1985). The mechanism of menadione-induced cytotoxicity is therefore thought to involve the creation of an "oxidative stress", following the depletion of GSH, during which key protein thiols are oxidized and the normal sequestration and removal of Ca^{2+} disrupted. This, in turn, may lead to a uncontrollable rise in cytosolic free Ca^{2+}, which subsequently results in cell death, possibly via destructive enzymes. Lipid peroxidation is unlikely to be involved in menadione-induced cytotoxicity, because menadione has such powerful antioxidant properties. It is important to note, however, that recent studies by Talcott et al. (1985) have shown that menadione will only inhibit microsomal lipid peroxidation in the presence of reducing equivalents, suggesting that

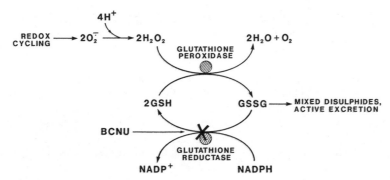

Figure 2. Inhibitory effect of BCNU on glutathione reductase and its consequent inhibition of H_2O_2 detoxication via glutathione peroxidase.

one or both of the reduced forms of menadione (semiquinone and hydroquinone) are the active antioxidants.

Further evidence for the importance of reactive oxygen species in menadione cytotoxicity has been obtained from recent studies in our laboratory on the combined effects of the chemotherapeutic drug BCNU and menadione on rat brain tumor cells. Pretreatment of the tumor cells with a low dose of BCNU (11.7 μM) has no significant effect on GSH levels, but almost completely inhibits glutathione reductase activity. This severely limits the cells' ability to detoxify H_2O_2 via glutathione peroxidase (Fig. 2). BCNU-pretreatment greatly enhances menadione cytotoxicity so that 4 logs of cell kill (i.e., the amount necessary to effect a "cure" of a solid brain tumor) can be achieved in a colony forming efficiency assay (CFE) of clonogenic tumor cells (C.G. Evans, D. Ross, and M.T. Smith, unpublished observations). The incidence of sister chromatid exchanges in the chromosomes of the tumor cells is also markedly increased. Thus, BCNU appears to greatly enhance the cytotoxic effects of H_2O_2 generated as a result of menadione redox cycling, a result which may be of clinical significance for brain tumor chemotherapy.

Diquat

The bipyridyl herbicides, paraquat and diquat are widely used throughout the world. Exposure to high levels of these compounds produces lung, liver and kidney injury. Both paraquat and diquat are readily reduced to their free radical forms which react rapidly with O_2. They are therefore potent redox cyclers, with diquat being much more effective than paraquat and approximately equivalent to menadione in effectiveness. Unlike menadione, however, diquat does not arylate thiols directly and is therefore an excellent model compound to study.

The importance of redox cycling in paraquat and diquat toxicity remains controversial. This is largely because one would expect the reactive oxygen species produced to stimulate lipid peroxidation, but some *in vivo* studies have failed to show significant evidences for this. Moreover, the potent antioxidant N,N'-diphenyl-*p*-phenylenediamine (DPPD) prevented *in vitro* paraquat stimulation of lipid peroxidation but failed to protect mice against lethal doses of paraquat (Shu et al., 1979).

We have recently used BCNU-pretreated isolated hepatocytes in an attempt to resolve this anomaly (Sandy et al., 1986). Diquat produces very rapid GSH depletion and toxicity in BCNU-pretreated hepatocytes and a concomitant rise in lipid peroxidation (Fig. 3). This lipid peroxidation could not be separated temporally from the toxic effects of diquat, unlike peroxidation produced by some other agents (Smith et al., 1982). The role of lipid peroxidation was therefore investigated by using three very potent antioxidants, namely DPPD, Trolox C and promethazine. All three antioxidants delayed, but did not prevent, diquat cytotoxicity (Fig. 3). These results are therefore in agreement with the *in vivo* data of Shu et al. (1979) in that antioxidants do not prevent the toxic effects of bipyridyl herbicides. Our studies do not, however, totally rule out a role for lipid peroxidation but suggest that it plays an augmentative, but non-essential role in bipyridyl herbicide toxicity.

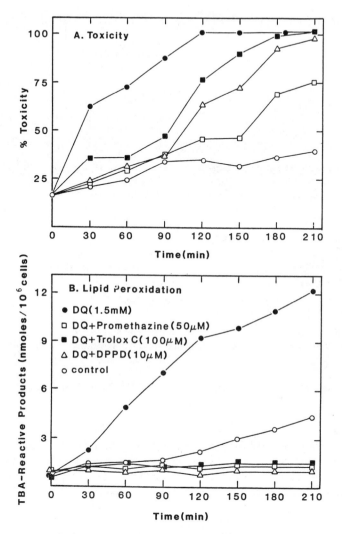

Figure 3. Effect of three different antioxidants on diquat-induced toxicity (A) and lipid peroxidation (B) in BCNU-treated hepatocytes. BCNU-treated hepatocytes were incubated at 10^6 cells/ml in Krebs-Henseleit buffer, pH 7.4, with either no additions (o), 1.5 mM diquat (●), 1.5 mM diquat + 50 μM promethazine (□), 1.5 mM diquat + 100 μM Trolox C (■) or 1.5 mM diquat + 10 μM DPPD (△). [Unpublished data of M.S. Sandy, P. Moldeus, D. Ross and M.T. Smith.]

The relationship between lipid peroxidation and chemical toxicity may therefore be summarized as follows:

(1) *Essential role* (e.g., BrCCl$_3$, CCl$_4$)

Chemical → Lipid Peroxidation → Cell Death

(2) *Augmentative, non-essential role* (e.g., Diquat, Paraquat)

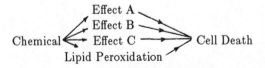

(3) *No role, resulting from toxicity rather than causing it* (e.g., Bromobenzene)

Chemical → Effect(s) → Cell Death → Lipid Peroxidation

GOALS FOR FUTURE RESEARCH

Reactive oxygen species are certainly mutagenic, but what types of damage do they produce in DNA at the molecular level and what is the prevalence of each form? What is the contribution of reactive oxygen-induced DNA damage to living organisms on a daily basis, i.e., does it form a major or minor component of the daily mutation rate? Also, what role does inflammation and phagocytosis play in causing DNA damage *in vivo*? New techniques aimed at measuring oxidative damage to DNA are required to answer these questions.

What role does the generation of reactive oxygen species and lipid peroxidation play in the metabolic activation of chemicals *in vivo*? Is this role significant for human carcinogenesis?

We know that reactive oxygen species can also be involved in the toxicity of various compounds and that lipid peroxidation may play at least some role in the toxicity of some compounds. However, what are the other toxic effects that reactive oxygen species can inflict on cells and what is their importance in producing the resultant toxicity?

Many fundamental questions therefore remain with respect to the role of reactive oxygen species in chemically-induced toxicity and carcinogenesis. One final quotation may be relevant on this matter: "I have yet to see any problem, however complicated, which, when you looked at it in the right way, did not become still more complicated." (Paul Anderson, *New Scientist*, 1969).

Acknowledgement. I thank the National Foundation for Cancer Research for their financial support. I am indebted to David Eastmond for his critical reading of the manuscript and to Michael Murphy for his expert secretarial assistance.

REFERENCES

Ames, B.N., and McCann, J., 1981, Validation of the salmonella test: a reply to Rinkus and Legator, *Cancer Res.*, 41:4192.

Bruyninckx, W.J., Mason, H.S., and Morse, S.A., 1978, Are physiological oxygen concentrations mutagenic?, *Nature*, 274:606.

Bus, J.S., and Gibson, J.E., 1984, Role of activated oxygen in chemical toxicity, in "Drug Metabolism and Drug Toxicity", J.R. Mitchell and M.G. Horning, eds., Raven Press, New York.

Chesis, P.L., Levin, D.E., Smith, M.T., Ernster, L., and Ames, B.N., 1984, Mutagenicity of quinones: pathways of metabolic activation and detoxification, *Proc. Natl. Acad. Sci. USA*, 81:1696.

Di Monte, D., Ross, D., Bellomo, G., Eklow, L., and Orrenius, S., 1984, Alterations in intracellular thiol homeostasis during the metabolism of menadione by isolated rat hepatocytes, *Arch. Biochem. Biophys.*, 235:334.

Fürstenberger, G., Kinzel, V., Schwarz, M., and Marks, F., 1985, Partial inversion of the initiation-promotion sequence of multistage tumorigenesis in the skin of NMRI mice, *Science*, 230:76.

Hochstein, P., 1983, Futile redox cycling: implications for oxygen radical toxicity, *Fund. Appl. Toxicol.*, 3:215.

Kappus, H., and Sies, H., 1981, Toxic drug effects associated with oxygen metabolism: redox cycling and lipid peroxidation, *Experientia*, 37:1233.

Kensler, T.W., and Trush, M.A., 1984, Role of oxygen radicals in tumor promotion, *Environmental Mutagenesis*, 6:593.

Levin, D.E., Hollstein, M., Christman, M.F., Schwiers, A., and Ames, B.N., A new *Salmonella* tester strain (TA102) with A·T base pairs at the site of mutation detects oxidative mutagens, *Proc. Natl. Acad. Sci. USA*, 79:7445.

Marnett, L.J., Hard, H.H., Hollstein, M.C., Levin, D.E., Esterbauer, H., and Ames, B.N., 1985, Naturally occurring carbonyl compounds are mutagens in *Salmonella* tester strain TA104, *Mutation Res.*, 148:25.

Orrenius, S., Thor, H., Di Monte, D., Bellomo, G., Nicotera, P., Ross, D., and Smith, M.T., 1985, Mechanisms of oxidative cell injury studied in intact cells, in "Microsomes and Drug Oxidations, 6th Int. Symp.", D.D. David et al., eds., Taylor and Francis, United Kingdom.

Sandy, M.S., Moldeus, P., Ross, D., and Smith, M.T., 1986, Role of redox cycling and lipid peroxidation in bypyridyl herbicide toxicity: studies with a compromised isolated hepatocyte model system, *Biochem. Pharmacol.*, in press.

Schwarz, M., Peres, G., Kunz, W., Fürstenberger, G., Kittstein, W., and Marks, F., 1984, On the role of superoxide anion radicals in skin tumour promotion, *Carcinogenesis*, 12:1663.

Shu, H., Talcott, R.E., Rice, S.A., and Wei, E.T., 1979, Lipid peroxidation and paraquat toxicity, *Biochem. Pharmacol.*, 28:327.

Sies, H., ed., "Oxidative Stress", Academic Press, London.

Smith, M.T., and Orrenius, S., 1984, Studies on drug metabolism and drug toxicity in isolated mammalian cells, in "Drug Metabolism and Drug Toxicity", M.D. Horning and J.R. Mitchell, eds., Raven Press, New York.

Smith, M.T., Thor, H., Hartzell, P., and Orrenius, S., 1982, The measurement of lipid peroxidation in isolated hepatocytes, *Biochem. Pharmacol.*, 31:19.

Smith, M.T., Evans, C.G., Thor, H., and Orrenius, S., 1985, Quinone-induced oxidative injury to cells and tissues, in "Oxidative Stress", H. Sies, ed., Academic Press, New York.

Talcott, R.E., Smith, M.T., and Giannini, D.D., 1985, Inhibition of microsomal lipid peroxidation by naphthoquinones: structure-activity relationships and possible mechanisms of action, *Arch. Biochem. Biophys.*, 241:88.

Thor, H., Smith, M.T., Hartzell, P., Bellomo, G., Jewell, S.A., and Orrenius, S., 1982, The metabolism of menadione in isolated hepatocytes. A study of the implications of oxidative stress in intact cells, *J. Biol. Chem.*, 257:12419.

Troll, W., and Wiesner, R., 1985, The role of oxygen radicals as a possible mechanism of tumor promotion, *Ann. Rev. Pharmacol. Toxicol.*, 25:509.

Trush, M.A., Seed, J.L., and Kensler, T.W., 1985, Oxidant-dependent metabolic activation of polycyclic aromatic hydrocarbons by phorbol ester-stimulated human polymorphonuclear leukocytes: possible link between inflammation and cancer, *Proc. Natl. Acad. Sci. USA*, 82:5194.

Weitzman, S.A., Weitberg, A.B., Clark, E.P., and Stossel, T.P., 1985, Phagocytes as carcinogens: malignant transformation produced by human neutrophils, *Science*, 227:1231.

SUGGESTIONS FOR FURTHER READING

Bus, J.S., and Gibson, J.E., 1984, Role of activated oxygen in chemical toxicity, in "Drug Metabolism and Drug Toxicity", J.R. Mitchell and M.G. Horning, eds., Raven Press, New York.

Kensler, T.W., and Trush, M.A., 1984, Role of oxygen radicals in tumor promotion, *Environmental Mutagenesis*, 6:593.

Sies, H., ed., 1985, "Oxidative Stress", Academic Press, London.

Slater, T.F., 1984, Free-radical mechanisms in tissue injury, *Biochem. J.*, 222:1.

Troll, W., and Wiesner, R., 1985, The role of oxygen radicals as a possible mechanism of tumor promotion, *Ann. Rev. Pharmacol. Toxicol.*, 25:509.

PRINCIPLES AND METHODS IN THE STUDY OF FREE RADICALS GENERATED IN THE METABOLISM AND CYTOTOXIC ACTION OF CHEMOTHERAPEUTIC AGENTS

J. William Lown

Department of Chemistry
University of Alberta
Edmonton, Alberta, Canada T6G 2G2

INTRODUCTION

Increasingly, the design, synthesis, selection and application of anticancer agents in chemotherapy is being replaced on a less empirical and more rational basis as a result of the detailed examination of mechanisms of action (Neidle and Waring, 1983; Pratt and Ruddon, 1979; Lown, 1982). It is becoming apparent that most agents exhibit several parallel pathways or effects on sensitive cell targets which include nucleic acids, certain proteins and membranes (Neidle and Waring 1982; Pryor, 1971).

The intermediacy and contribution of free radical species in the metabolism and cytotoxic action of many agents, including anticancer drugs, has been recognized relatively recently (Pryor, 1971). The significance of the findings on mechanism of action has been increased by the concomitant recognition of a number of factors that may contribute to the selective activation of anticancer drugs and the susceptibility of certain tissues, including neoplastic tissue, to oxidative damage induced by free radicals. This chapter will attempt to discuss the techniques used to detect, identify, and quantify such species, to elucidate molecular mechanisms of action, and the application of this information to drug design.

Classes of Anticancer Agents in Which Free Radicals Have Been Implicated in Their Mode of Action

Anticancer agents fall into six broad categories: (i) alkylating agents, (ii) antibiotics, (iii) antimetabolites, (iv) mitotic inhibitors, (v) steroidol hormones, and (vi) miscellaneous drugs, e.g., *cis*-platinum complexes (Neidle and Waring, 1983). Of these general classes of agents, examples involving free radical species have been found frequently in type (ii) and include (a) quinones, (b) peptides and glycopeptides, and (c) certain alkaloids.

FREE RADICAL PATHWAYS IN DRUG METABOLISM AND ACTION

Many of these types of drugs undergo chemical modification as a result of metabolism, so that in many cases more than one species is involved in cellular damage (Neidle and Waring, 1983). Metabolic conversion may involve a number of different processes, including:

Enzymatic Reduction

Drugs of the quinone class are subject to one electron reduction by redox flavoenzymes in which the electron donor is often the flavin cofactor. The microsomal reduction of quinones to semiquinones is catalyzed by NADPH-cytochrome c (P-450) reductase in which one-electron transfer to these electron acceptors is obligatory. Several different types of enzymes are capable of one electron reduction of quinones or quinone-imines, including NADH-dependent dehydrogenase, xanthine dehydrogenase, xanthine oxidase, lipamide dehydrogenase and ferredoxin-NADP reductase. Hitherto unidentified enzymes capable of reductive activation of the important anticancer agents daunorubicin and dubbed "daunorubicin reductases" presumably belong to one of these categories.

Enzymatic Oxidation

One electron oxidation of anticancer agents containing hydroquinone or phenol moieties may occur by, for example, peroxidases, ceruloplasmin, and laccase which employ H_2O_2, O_2, and O_2 respectively as cofactors. Such enzymes may be involved in the oxidative metabolism of the pyrrolo [1,4] benzodiazepine antibiotics, anthramycin and sibiromycin, and in the generation of free radicals from podophyllotoxin. It is conceivable that oxidative metabolism and oxygen radical generation are synergistic in some cases since $O_2^-\cdot$ induces aryl hydrocarbon hydroxylase.

Metal Ion Sequestration or Exchange

Certain antitumor agents have a cofactor requirement for specific metal ions in their cytotoxic function. Frequently the resulting metal ion complexes are involved in the generation of free radicals or other paramagnetic species (Neidle and Waring, 1983; Lown, 1982).

Metal ions have been implicated in the mode of action of:

Streptonigrin (Cu^{++}, Zn^{++}, Mn^{++})	(1)
Bleomycin (Cu^{++}, Fe^{++}, Co^{++})	(2)
Tallysomycin (Fe^{++} plus second metal ion Zn^{++} or Mg^{++})	(3)
Doxorubicin (Fe^{++})	(4)
Aureolic acid antibiotics (Mg^{++})	(5)

Streptonigrin, the aminoquinone antibiotic, which has been shown to function by generation of O_2^-· and OH· radicals, has specific metal ion requirements (Lown, 1982). Its unique bipyridyl containing structure sequesters metal ions Zn^{++} or Mn^{++} in two chemically distinct ways, depending on the hardness or softness of the metal ion (Fig. 1). The nature of the resulting charged metal complex in turn controls both the ease of reduction-reoxidation in the generation of free radicals (Cu > Mn >> Zn) and the binding of the antibiotic to DNA (Lown, 1982).

The glycopeptide antibiotics bleomycin and tallysomycin also show specific metal ion sequestration requirements (Hecht, 1979). Bleomycin binds to DNA and chelates adventitious Fe(II) to initiate the redox cycle that produces oxygen-dependent scission (Hecht, 1979). Tallysomycin exhibits a similar specific requirement for iron to express its cytotoxic properties but also binds a second metal ion (Zn^{++} or Mg^{++}) at the talose and β-lysinespermidine moieties. The latter sequestered metal ion appears to be involved with the binding of the antibiotic to a particular sequence of DNA rather than in the redox activity.

Preferential Protonation in Tumor Tissue

Tumor tissue generally shows a slightly lower average pH than normal tissue, owing to an increased rate of glycolysis and concomitant increased production at lactic acid. This has an indirect effect on free radical generation and disposition in the case of drugs susceptible to acid catalyzed covalent attachment to cell targets such as DNA. Consequently, drugs which are specifically activated by protonation may exhibit a slight preferential effect.

Detection of Free Radical Intermediates in Drug Action and the Lesions They Cause in Macromolecules

Free radical intermediates may, provided certain conditions are fulfilled, be detected *in vivo* or *in vitro* by ESR spectroscopy (McLaughlin, 1972; Wertz, 1972). The

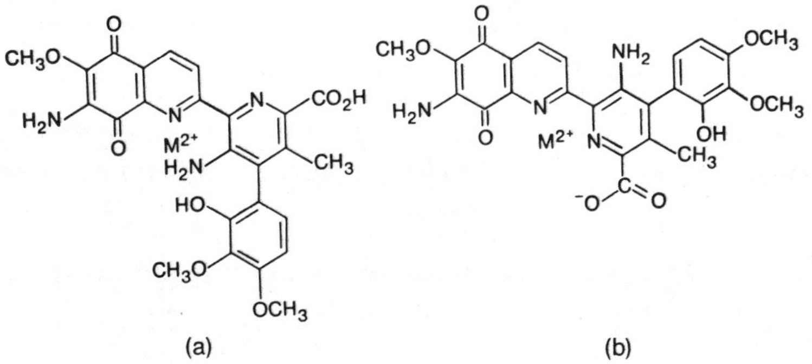

Figure 1. Two different metal chelation complexes formed by antibiotic streptonigrin.

169

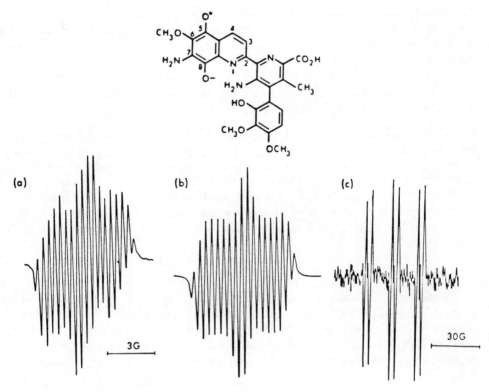

Figure 2. (a) ESR spectrum of streptonigrin semiqunone; (b) computer simulated spectrum; (c) ESR spectrum of PBN.OH spin trapped radical form streptonigrin semiquinones.

latter technique permits the determination of both the structure of a free radical and its concentration. To permit ESR detection of free radicals, the species must be sufficiently long-lived or their equilibrium concentration sufficiently high. The concentration of components with which they react (for example, oxygen) must also be sufficiently low. Moreover, aqueous solutions, frequently used of necessity with nucleic acids or proteins, cause additional detection problems for ESR owing to dielectric loss. Attempts should be made to establish the identity of the ESR signals of free radicals obtained *in vitro* with those obtained from the correspnding drug *in vivo*. This is often problematic owing to the poor resolution (or lack of it) frequently encountered in the latter case as illustrated in Fig. 2. Another useful ESR technique is to use the line broadening caused by paramagnetic O_2 as a probe for O_2 concentration and, indirectly, for the detection of $^1O_2$ (vide infra) [see Fig. 3].

The fact that many DNA damaging anticancer agents cause two or more types of lesions simultaneously poses severe analytical problems. Moreover, special problems

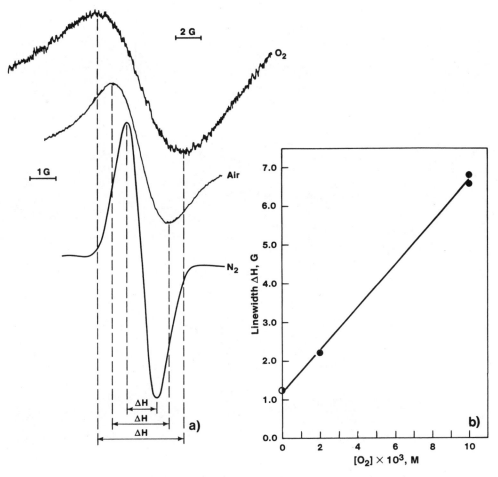

Figure 3. (a) Width (ΔH) of the ESR line ($m_1 = +1$) of TEMPO in oxygen, air and nitrogen saturated ethanol. (b) Dependence of ΔH of the ESR line ($m_1 = +1$) of TEMPO on oxygen concentration in ethanolic solution.

attend the detection of DNA lesions in intact cells or in purified native DNAs, owing to their fragility and sensitivity to shear forces. The techniques employed include: hyperchromicity and absorption spectral changes, gel electrophoresis, radiolabelling methods, sedimentation and alkaline elution. The latter method, as well as certain fluorescence assays, provides quite detailed information on the different kinds of lesions (Lown, 1982). For example, an assay suitable for the detection of O_2-mediated DNA scission is illustrated in Fig. 4. The principal advantage of this assay is that by adding inhibitors, such as the cell protective enzymes superoxide dismutase, catalase or glutathione peroxidase, the chemical mechanism of the scission process may thereby be investigated (Lown, 1982; Lown, 1983). Many such assays may be devised (Lown, 1982).

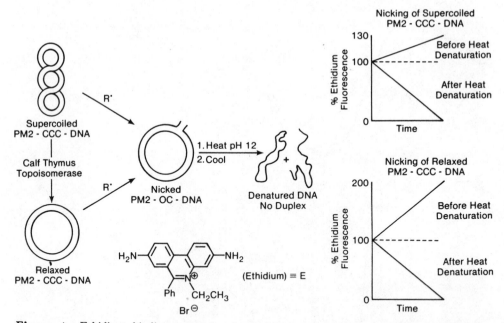

Figure 4. Ethidium binding assay for determining single strand breakage of supercoiled covalently closed circular DNA. The release of constraints upon nicking the DNA to give the open circular form (O.C.) permits more DNA to intercalate and give a 30-38% rise in fluorescence, depending on the superhelical density. Denaturation at pH 12 prevents duplex formation, and fluorescence falls ultimately to zero.

Complementary assays that are useful in the study of O_2-mediated drug action include:

(a) *Augmentation of Oxygen Consumption by Hepatic Microsomes:Application to Redox-Active Drugs.* This is a characteristic property of many redox active drugs, and its utility for investigating the origin of the cardiotoxicity of anthracyclines is illustrated later. The procedure in summary is:

1. Exact microsomes from livers of male Sprague-Dawley rats.

2. Suspend frozen microsomes in Triton N-101 and 0.1 M potassium phosphate buffer pH 7.5.

3. Determine protein levels using dye-binding procedure with human albumin as protein standard.

4. Aliquot of microsomes, NADPH and drug equilibrated.

5. Determine endogenous O_2 consumption rate using Clark electrode and biological O_2 monitor.

6. Couple output voltage of O_2 monitor to an analog-to-digital multimeter for recording data.

(b) *Electrochemical Study of Redox-Active Drugs and Intermediates*. Applications:

1. Polarography — Redox potentials.
2. Cyclic Voltammetry — quantity, rate, and extent of reoxidizability.
3. Controlled potential coulombetry — determine number of electrons involved in each process.
4. Rapid sweep voltammetry — determine drug:DNA binding constants.

(c) *Gel Electrophoresis*: especially agarose gels for detecting single and double strand-O_2 mediated scission of circular DNAs.

Study of the Chemical Reactions of Oxygen Species Implicated in the Cytotoxic Action of Chemotherapeutic Agents

Free radicals have been shown to cause damage to nucleic acids polysaccharides, membrane lipids, free aminoacids, enzymes, as well as structural, receptor and transport proteins of these processes (Jayaishi and Asada, 1977). Damage to DNA and lipids have been investigated most thoroughly (Hayaishi and Asada, 1977). We will define "damage" to DNA or to other cell targets as any alteration involving nonbonded interactions, chemical bond formation, or permanent or transient bond breakage that affect normal cellular functions. Such damage is lethal however only when the critical changes persist for longer than the period of time that the cells can survive the lesion. "Permanent" damage implies that lesions are not repaired, replaced by biosynthesis or uptake, or by-passed through alternative metabolic pathways (Simic, 1985; Lown, 1983b).

TYPES OF CHEMICAL LESIONS IN CELLULAR TARGETS

Nucleic Acids

The different types of chemical lesions often induced in DNA by antitumor agents (many of which are O_2-mediated) are summarized in Fig. 5. Several of these occurring simultaneously pose severe analytical problems which may often be solved with ethidium binding assays (Lown, 1982).

Techniques Useful in Study of DNA Active Drugs

The interdependence of parallel modes of action commonly encountered with anti-cancer drugs often requires examination of several different aspects of drug action. The following techniques are useful in this regard:

Static Aspects

1. Electron microscopy for DNA topology
2. X-ray diffraction
3. Viscosity measurements, topoisomerase I unwinding angles
4. 2D-high-field NMR: especially NOESY for through-space interactions

173

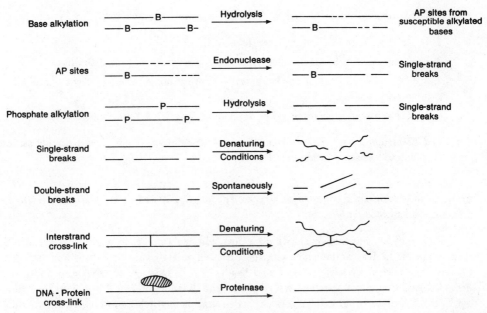

Figure 5. Different types of chemical lesions produced in DNA by antitumour agents.

Dynamic Aspects: mechanisms

1. Agarose gel electrophoresis to detect strand breakage and cross-links
2. Alkaline elution to detect strand breaks and cross-links
3. Electrochemistry of redox active drugs
4. Kinetics of drug dissociation from DNA complexes
5. Ethidium binding assays with selected repair enzymes to pinpoint specific lesions

Membrane Lipids

Cellular membranes are especially sensitive to oxidation reactions mediated by radicals because of the unsaturated nature of the fatty acid chains of phospholipids of cytomembranes and because allylic hydrogens may be readily abstracted (Hayaishi and Asada, 1977). Another factor is that molecular oxygen and its derivatives are 7-8 times more soluble in hydrophobic lipid regions than in hydrophilic areas. Therefore, it may be expected that leakage of reactive oxygen species from the electron transport systems associated with mitochondria or the endoplasmic reticulum, could initiate lipid peroxidation of membranes (Hayaishi and Asada, 1977). Under normal circumstances, the enzymatic surveillance system and antioxidant cellular defence mechanisms (such as α-tocopherol) are adequate to control this low level damage. A significant increase in radical generation due to xenobiotics would result in depletion of this and other antioxidants or cause inactivation in certain of the surveillance enzymes. The latter may additionally be depleted in neoplastic tissue and in certain organs. These factors are relevant to the discussion of clinical toxicities (vide infra).

Proteins and Enzymes

Anticancer drugs, especially those of the intercalator class, induce DNA-protein cross-links that are often associated with repair processes (Pratt and Ruddon, 1979). Nucleophilic sites in proteins and enzymes ($-SH,-NH_2$) are often implicated in attack by electrophiles from anticancer drugs, including Michael addition. These types of reactions are often associated with the property of inhibition of microtubulin formation in mitotic inhibitors such as maytansine and vincristine (Neidle and Waring, 1983). Cross-link formation within proteins has been observed as a secondary result of initial free radical induced lipid peroxidation. However, recent evidence on DNA-protein cross-links points, in certain cases, to selective inhibition of topoisomerase II (Simic, 1985).

MECHANISMS OF REACTION OF OXYGEN SPECIES FROM ANTITUMOR AGENTS WITH CELLULAR MACROMOLECULES — CLINICAL APPLICATIONS AND APPROACHES TO NEW DRUG DESIGN

Although many compounds, both natural and synthetic, exhibit anticancer activity, relatively few survive into clinical utility because of unacceptable toxicities (Montgomery et al., 1979; Cassady and Douros, 1980). A recent encouraging trend has been to attempt to elucidate the molecular origin of toxicities and, by appropriate chemical modification, to improve the therapeutic index. Illustrative examples of current interest include the anthracyclines and glycopeptide antibiotics.

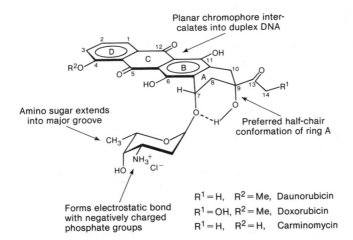

Figure 6. Structures and stereochemistry of daunorubicin, doxorubicin, and carminomycin, showing points of attachment to duplex DNA.

Anthracyclines

The anthracyclines (Fig. 6) arguably comprise the most important, and certainly the most widely prescribed, clinically useful anticancer agents (Arcamone, 1981). The anthracyclines exhibit three principal cellular effects: (i) inhibition of nucleic acid metabolism, (ii) redox reactions and the resulting interference with electron transport and cell respiration, and (iii) initiation of lipid peroxidation and consequent disruption of cellular phospholipid membranes. Free radicals have been implicated in all three processes but are not necessarily dominant in all three effects.

The effects of the facile *in vivo* redox reactions of the quinone moiety of anthracyclines is apparent on several cell functions. Thus doxorubicin and daunorubicin exhibit significant inhibition of respiration of isolated mitochondria of normal and tumor cells (Arcamone, 1981). Doxorubicin, carminomycin, daunorubicin and doxorubicin-14-octanoate inhibit NADH-oxidase and succinoxidase, the cell respiratory chain enzymes that require coenzyme Q_{10} as a cofactor (Arcamone, 1981). The cellular effects are in accord with the known involvement of redox reaction at the quinone moieties of anthracyclines in their metabolism (Fig. 7). Anthracyclines in the presence of rat liver microsomes or cytochrome P-450 reductase led to O_2 consumption and membrane lipid peroxidation. Subsequent studies indicated this to be related to the onset of cardiotoxicity (Lown, 1983a).

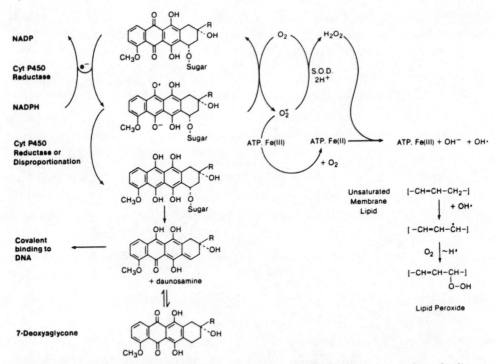

Figure 7. Cytochrome P450 reductase reduction of anthracycline chromophore leading to deglycosidation and formation of the 7-deoxyaglycone and the concomitant interaction with molecular oxygen leading to reactive oxygen species and lipid peroxidation.

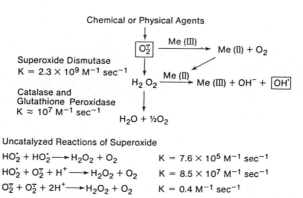

Figure 8. Scheme depicting the enzymatic surveillance defense mechanism against cellular oxygen toxicity.

Morphological evidence for damage in rat heart tissue is thus interpreted in terms of the production of a flux of reactive oxygen species from the activated anthracycline which overwhelms the cellular enzymatic surveillance system (Fig. 8).

Certain cellular protective mechanisms have evolved against oxygen toxicity as a result of the introduction of molecular oxygen in the atmosphere (Hayaishi and Asada, 1977). The first line of defense is the enzyme class of superoxide dismutases. The latter, which occur in both Cu/Zn and Mn forms, dismutate O_2^- to H_2O_2 and O_2 at rates approaching the diffusion limit (Hayaishi and Asada, 1977). These enzymes work in concert with catalase and glutathione peroxidase which remove H_2O_2 (Fig. 8). In this case these cellular protective enzymes prevent the formation of OH· radical, against which there is no known enzymatic defense, by removing its precursors.

Under normal circumstances this enzymatic surveillance system is efficient in removing reactive oxygen species. However, a number of factors are becoming recognized which may promote the susceptibility of certain tissues to oxidative insult (Lown, 1983a). These include (i) apparent variation in the levels of the enzymes in different organs, (ii) differences in the enzyme levels in normal and neoplastic tissue, (iii) local concentration of oxygen-radical generating drugs to overwhelm the protective enzymes, and (iv) selective inhibition of catalase by O_2^-· or of individual protective enzymes by a given drug.

Correlations are observed between *in vivo* cardiotoxicity of a group of anthracyclines and their ability to augment hepatic microsomal oxygen consumption and produce OH· radical mediated damage to DNA. An analysis of the structural factors in the drug chromophore and in the sugar moiety controlling the generation of reactive oxygen species using these correlations permitted a separation of antileukemic and cardiotoxic effects (Lown, 1983a) [Fig. 9]. In the group of derivatives of daunorubicin and doxorubicin, a twenty-fold decrease in cardiotoxicity was achieved with no sacrifice in antileukemic properties (Lown, 1983a). A 5-imino substitution substantially decreases redox activity compared with the parent anthracycline in accord with the lack of

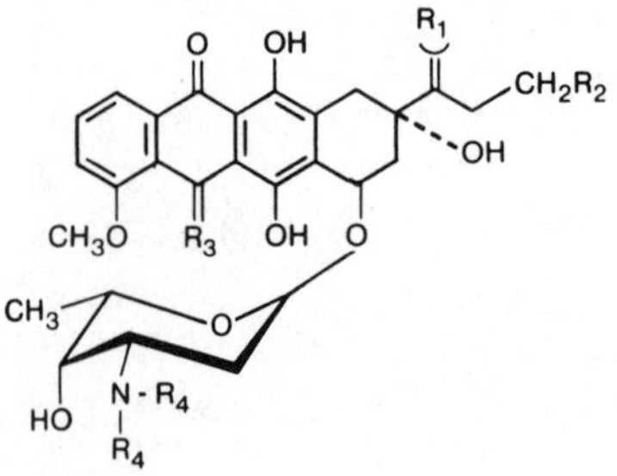

$1 \quad R_1 = 0, \quad R_2 = H, \quad R_3 = 0, \quad R_4 = CH_3$
$2 \quad R_1 = 0, \quad R_2 = OH, \quad R_3 = 0, \quad R_4 = H$
$3 \quad R_1 = 0, \quad R_2 = H, \quad R_3 = 0, \quad R_4 = H$
$4 \quad R_1 = 0, \quad R_2 = OH, \quad R_3 = 0, \quad R_4 = CH_3$
$5 \quad R_1 = 0, \quad R_2 = H, \quad R_3 = NH, \quad R_4 = H$
$6 \quad R_1 = 0, \quad R_2 = H, \quad R_3 = 0, \quad R_4 = CH_2Ph$

Anthracycline	$\Delta Tm°C$	Activity against P388 Leukemia mice % T/C	Cardiotoxicity (rats) min. cum. dose (mg/kg)	O$_2$ consumption rat liver microsomes % activity Adriamycin	E° (quinone) Volts	Relative O$_2^{\bar{}}$ Generation (90 μM Drug)
1	17.5	214	6	145	−0.65	90
2 Adriamycin	13.4	197	11	100	−0.64	88
3 Daunorubicin	11.2	160	14	109	−0.64	90
4	15.3	164	11 - 16	108	−0.64	90
5	6.25	170	64	8.2	−0.70	55
6	1.4	259	125	0.5	−0.65	10

Figure 9. Aglycone modified anthracyclines—effects on oxidizing properties and cardiotoxicity.

reductive metabolism and the significantly lower cardiotoxicity (Lown, 1983a). 5-Imino-daunorubicin is now in preclinical toxicity trials as a result.

Less Cardiotoxic Clinical Alternatives to Anthracyclines - Mitoxantrone and Anthrapyrazoles

Because the severe risk of cardiotoxicity of doxorubicin (Arcamone, 1980) continues to impose severe restrictions on the clinical utility of this important agent, efforts continue to develop less cardiotoxic alternatives.

New synthetic agents which are structurally related to anthracyclines offer an apparent clinical advantage in lower cardiotoxicity and with concomitant suppression of *in vitro* and *in vivo* generation of reactive oxygen species.

Mitoxantrone and Ametantrone

A promising agent which bears a superficial structural resemblance to the anthracyclines is mitoxantrone (Fig. 10) and the related ametantrone (Smith, 1983). Biochemical evidence suggests that the cytotoxic action of mitoxantrone is due, in part, to intercalative binding to duplex DNA. The latter phenomenon has been demonstrated by electron microscopy. The drug shows a preference for G.C base pairs with a binding constant with calf thymus DNA in 0.1 M Na$^+$, pH = 7 of 6 x 10^6 M^{-1}. The rate constant for the SDS driven dissociation of mitoxantrone from the DNA complex at 20°C is 1.3 sec^{-1}. The unwinding angle of mitoxantrone is 17.5° (Lown, et al., 1985).

The relatively low *in vivo* redox activity of mitoxantrone is reflected in the polarographic redox potential of –0.775v compared with –0.625v for doxorubicin. Similarly, the augmentation of hepatic microsomal O_2 uptake is 16.5% that of doxorubicin. Thus the relatively lower incidence of cardiotoxicity found for mitoxantrone (Smith, 1983) may reflect its diminished capacity to engage in redox reactions *in vivo* with the consequent decreased production of reactive oxygen species.

Mitoxantrone is a deep blue color with absorption maxima at 664 and 612 nm. This color liability from the clinical standpoint has an unexpected effect in that the compound can photosensitize with visible light, albeit weakly, the production of reactive singlet oxygen $^1O_2$ (Fig. 11). This effect is more pronounced with certain of the structurally related anthrapyrazoles (q.v.).

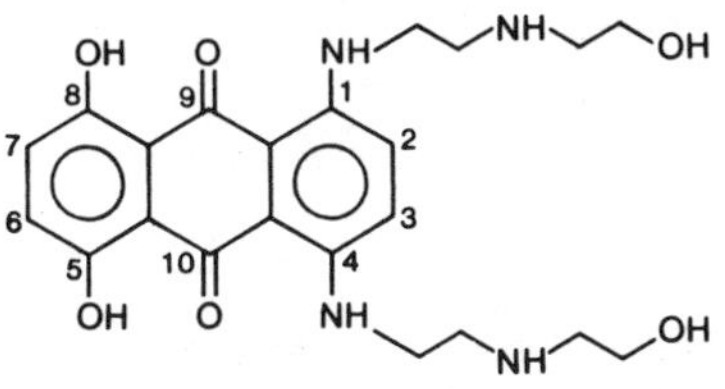

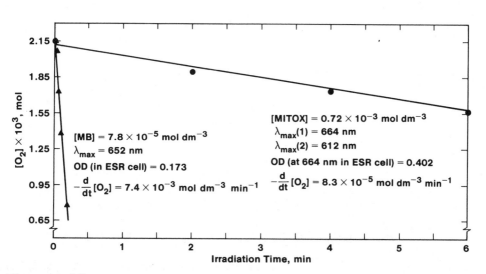

Figure 10. Structure and numbering of mitoxantrone.

[MB] = 7.8×10^{-5} mol dm^{-3}
λ_{max} = 652 nm
OD (in ESR cell) = 0.173
$-\dfrac{d}{dt}[O_2]$ = 7.4×10^{-3} mol dm^{-3} min^{-1}

[MITOX] = 0.72×10^{-3} mol dm^{-3}
$\lambda_{max}(1)$ = 664 nm
$\lambda_{max}(2)$ = 612 nm
OD (at 664 nm in ESR cell) = 0.402
$-\dfrac{d}{dt}[O_2]$ = 8.3×10^{-5} mol dm^{-3} min^{-1}

[O_2] × 10^3, mol

Irradiation Time, min

Figure 11. Mitoxantrone and methylene blue photosensitized oxygen consumption during irradiation with visible light in methanol with dimethylfuran as $^1O_2$ trap.

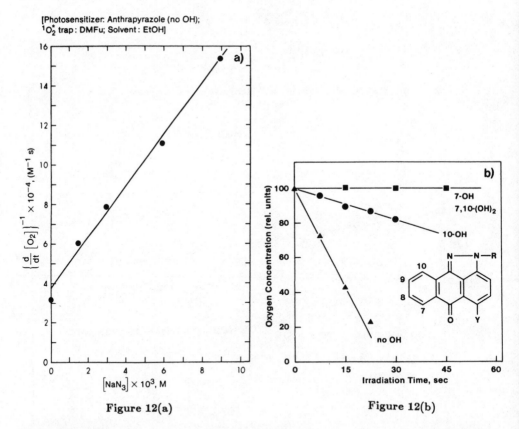

Compound	Z	R	Y
1	—	$(CH_2)_2NH(CH_2)_2OH$	$NH(CH_2)_2NH(CH_2)_2OH$
2	7,10 $(OH)_2$	$(CH_2)_2NH(CH_2)_2OH$	$NH(CH_2)_2NH(CH_2)_2OH$
3	7(OH)	$(CH_2)_2NH(CH_2)_2OH$	$NH(CH_2)_2NH(CH_2)_2OH$
4	10(OH)	$(CH_2)_2NH(CH_2)_2OH$	$NH(CH_2)_2NH(CH_2)_2OH$
5	7,10$(OH)_2$	$(CH_2)_2OH$	$NH(CH_2)_2NH(CH_2)_2OH$
6	7,10$(OCH_2Ph)_2$	$(CH_2)_2NCH_2Ph(CH_2)_2OH$	$NH(CH_2)_2CH_3$

Figure 12(a)

Figure 12(b)

Figure 12. (a) Dependence of the reciprocal of the rate of oxygen consumption by anthrapyrazole on the NaN_3 concentration. (b) Effect of chromophore substituents on anthrapyrazole-photosensitized oxygen consumption.

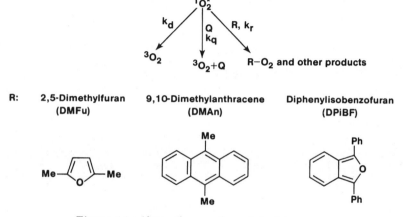

$^1D \xrightarrow{h\nu} {}^1D^*$	excitation
$^1D^* \xrightarrow{k_f} {}^1D$	fluorescence
$^1D^* \xrightarrow{k_{IC}} {}^1D$	internal conversion
$^1D^* \xrightarrow{k_{isc}} {}^3D^*$	intersystem-crossing
$^3D^* \xrightarrow{k_{ph}} {}^1D$	phosphorescence
$^3D^* + {}^3O_2 \xrightarrow{k_O} {}^1D + {}^1O_2^*$	Oxygen quenching via energy transfer

Figure 13. Schematic representation of photochemical generation of singlet oxygen.

1O2*
kd
Q
kq
R, kr
3O2
3O2+Q
R−O2 and other products
R:
2,5-Dimethylfuran
(DMFu)
9,10-Dimethylanthracene
(DMAn)
Diphenylisobenzofuran
(DPiBF)
Me
Me
Me
Me
Ph
O
Ph
O

Figure 14. Alternative reactions of singlet oxygen.

Anthrapyrazoles

Chromophore modification of the anthracenediones related to mitoxantrone in an attempt to provide agents with diminished or no cardiotoxicity resulted in a novel class of DNA binders, the anthrapyrazoles. The anthrapyrazoles (Showalter et al., 1983) bind strongly to DNA, are selective and potent inhibitors of DNA synthesis, and cause the formation of single-strand breaks in DNA. They also produce far less (20 to 200-fold) superoxide dismutase sensitive O_2 consumption than doxorubicin in the rat liver microsome system, a property that may be indicative of reduced cardiotoxicity. This result is in accord with their polarographic properties in which the anthrapyrazoles show a much greater resistance to reduction ($E'_{1/2} = -0.983$ to -1.089v) relative to doxorubicin ($E'_{1/2} = -0.625$v) and mitoxantrone ($E'_{1/2} = -0.775$v). While the color liability has been removed, modification of the chromophore in this case renders certain of these agents photosensitizing to visible light. By using an ESR probe for O_2 consumption described above, and in conjunction with a specific $^1O_2$ quencher sodium azide, it was demonstrated that certain anthrapyrazoles photosensitize the generation of $^1O_2$. The photochemical generation of $^1O_2$ is summarized in Fig. 13, while the characteristic reactions of $^1O_2$ with traps whereby it is detected are shown in Fig. 14. Fig 12(b) illustrates some of the structural parameters contributing to photosensitization in this series. Intramolecular hydrogen-bonding appears to be significant. While the particular biological consequences of the photodynamic effect (Hayaishi and Asada, 1977) in the case of the anthrapyrazoles are yet to be demonstrated, some effects of $^1O_2$ that may be anticipated are illustrated in Fig. 15.

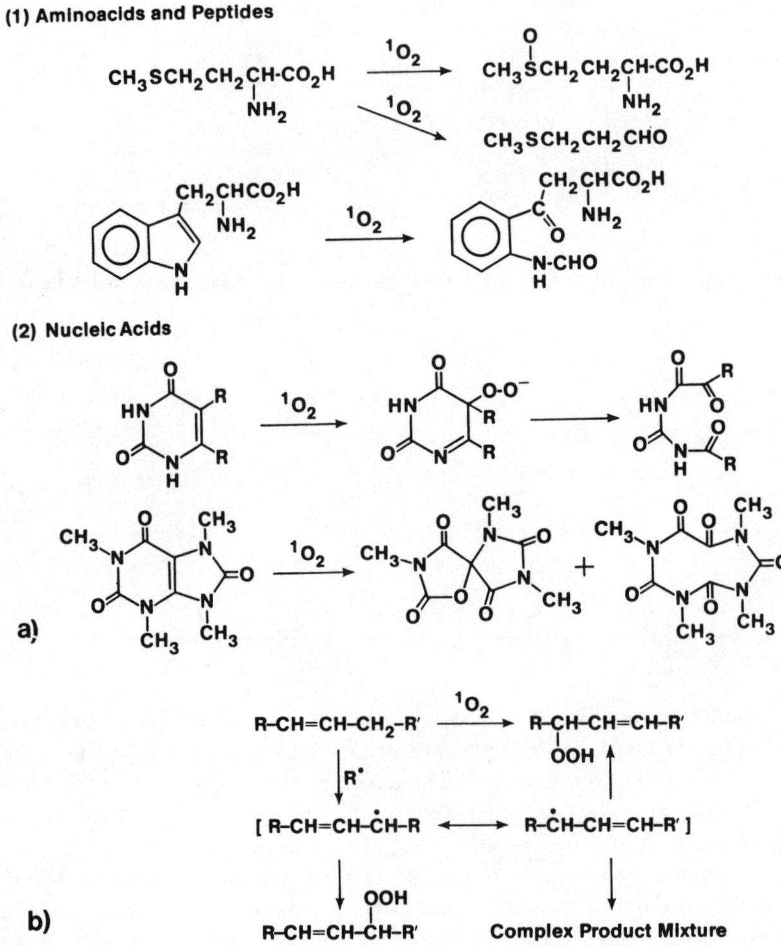

Figure 15. Biological effects of singlet oxygen on aminoacids, peptides, nucleic acids and membrane lipids.

A2 : R=NHCH2CH2CH2S+(CH3)CH3
B2 : R=NHCH2CH2CH2CH2NHC(=NH)NH2

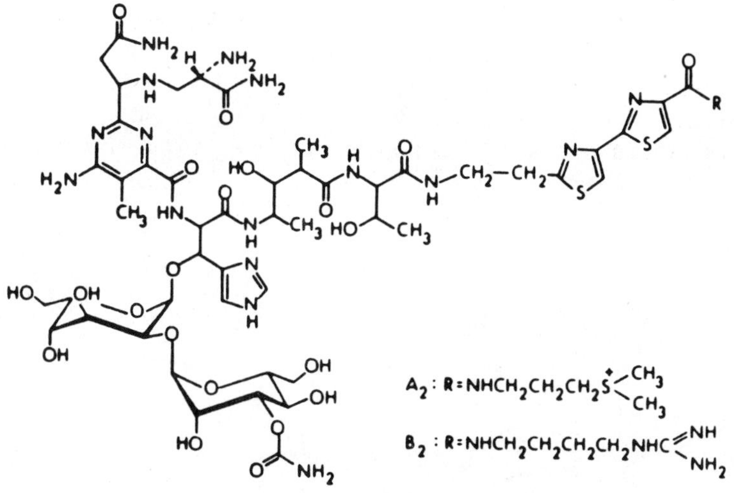

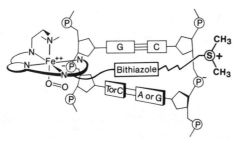

Figure 16. Structure of bleomycin A_2 and B_2.

G
C
Fe++
Bithiazole
CH3
CH3
TorC
A or G

Figure 17. Representation of the suggested mode of action of the antitumour antibiotic bleomycin in site specific binding to DNA and strand scission initiated by stereochemically directed hydrogen atom abstraction at the deoxy sugar moiety.

Bleomycin and its Pulmonary Toxicity

A somewhat different but related example of an organ toxicity which may also have its origin in local free radical generation is the case of bleomycin (Hecht, 1979). This relates to the severe clinical problem of pulmonary toxicity (Hecht, 1979). Bleomycin accumulates in the lung, or rather is not inactivated in this organ because of the apparently low level of the enzyme bleomycin hydrolase (Hecht, 1979). At the same time, the drug, which requires both a reducing cofactor and oxygen in its cytotoxic action, is naturally exposed to a high concentration of oxygen.

While the cytotoxic action of bleomycin on DNA involves a glycopeptide-iron-oxygen-DNA complex which may not generate free oxygen radicals, nevertheless in the absence of DNA, oxygen free radicals are generated which can cause damage including

lung membrane lipid peroxidation. The implication that this may represent the underlying cause of the pulmonary toxicity of bleomycin is plausible. Efforts to ameliorate the pulmonary toxicity by changes in the structure of bleomycin have not led to marked improvements. Given the evident requirement for an oxygen and free-radical mediated cytotoxic action, this approach seems unpromising unless bleomycins can be developed which do not accumulate in the lung.

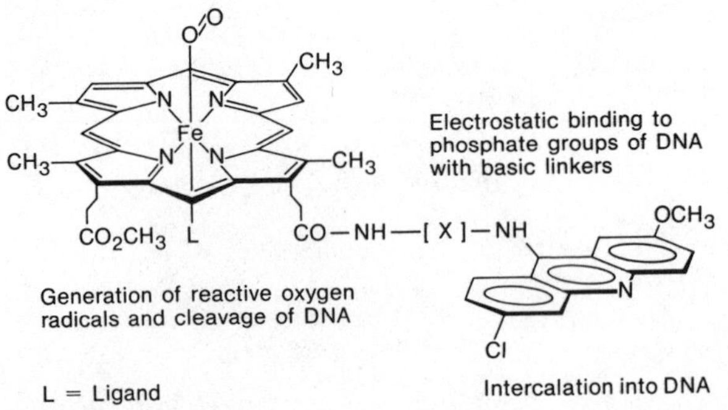

Figure 18. Representation of a possible mode of action of anticancer hemin-acridine bleomycin models in intercalative binding to duplex DNA and iron-oxygen mediated strand scission.

One development that may be useful in this regard is the design and synthesis of active functional bleomycin models based on hemin-acridines (Lown et al., 1984) Fig. 18. These agents mimic some of the characteristic cellular properties of bleomycin in (i) binding to duplex DNA, (ii) forming an Fe(III)O$_2$ complex, (iii) producing efficient strand breakage, (iv) exhibiting preferential cytotoxicity towards *E. coli* under aerobic conditions, and (v) showing antileukemic properties and reproducing the resistance properties against B16 melanoma (Lown et al., 1984). In contrast to the glycopeptides, the porphyrin or hemin moiety is not known particularly to accumulate in the lung but rather to concentrate in tumor tissue. It is therefore possible that further development of these agents may provide a less toxic class of anticancer drugs.

CONCLUSION AND PROSPECTS

In summary, the recent recognition of the significance of free radical mediated reactions in biological systems is having a pronounced effect in medicine. Such reactive intermediates appear to be important both in the development of certain disease states and increasingly in their treatment. An encouraging trend is the attempt to interpret some specific organ toxicities caused by chemotherapeutic agents in terms of the generation of free radical species and the incorporation of this rapidly accumulating information into more effective drug design.

REFERENCES AND SUGGESTED READING

Arcamone, F., 1981, "Doxorubicin Anticancer Antibiotics," Academic Press, New York.

Cassady, J.M., and Douros, J.D., eds., 1980, "Anticancer Agents Based on Natural Product Models," Academic Press, New York.

Hayaishi, O., and Asada, K., eds., 1977, "Biochemical and Medical Aspects of Active Oxygen," University Park Press, Baltimore.

Hecht, S.M., ed., 1979, "Bleomycin, Chemical, Biochemical and Biological Aspects," Springer-Verlag, New York.

Lown, J.W., 1982, Newer approaches to the study of the mechanism of action of antitumor antibiotics, *Accts. Chem. Res.*, 15:381.

Lown, J.W., 1983, Mechanism of action of quinone antibiotics, *Molec. Cell. Biochem.*, 55:17.

Lown, J.W., 1983b, The chemistry of DNA damage by antitumor drugs, in "Molecular Aspects of Anticancer Drug Action," S. Neidle and M.J. Waring, eds., Macmillan, pp. 283-314.

Lown, J.W., Plenkiewicz, J., Ong., C.W., Joshua, A.V., McGovren, J.P., and Hanka, L.J., 1984, Models for bleomycin antitumor antibiotics, *Proc. 9th IUPHAR Congress*, London.

Lown, J.W., Morgan, A.R., Yen, S.F., Wang, Y.H., and Wilson, W.D., 1985, Characteristics of the binding of the anticancer agents mitoxantrone, ametantrone and related structures to DNA, *Biochemistry*, 24:4028.

McLaughlin, J., 1972, "EPR", Oxford.

Montgomery, J.A., Johnston, T.P., and Shealy, Y.F., 1979, Drugs for neoplastic diseases, in "Burger's Medicinal Chemistry," 4th Edition, Part II, p. 595, M.E. Wolff, ed., Wiley, New York.

Neidle, S., and Waring, M.J., eds., 1983, "Molecular Aspects of Anticancer Drug Action," Topics in Molecular and Structural Biology 3, Macmillan, London.

Pratt, W.B., and Ruddon, R.W., 1979, "The Anticancer Drugs," Oxford, New York.

Pryor, W.B., ed., 1971, "Free Radicals in Biology," Vol. 1, Academic Press, New York.

Showalter, H.D.H., Johnson, J.L., Werbel, L.M., Leopold, W.R., Jackson, R.C., and Elslager, F.E., 1984, 5-[(Aminoalkyl)amino]-substituted anthra[1,9-cd]pyrazol-6[2H]-ones as novel anticancer agents, *J. Med. Chem.*, 27:253.

Simic, M., ed., 1985, Proceedings of Conference, "Mechanisms of DNA Damage and Repair," Gaithersburg, MD, June 2-7.

Smith, I.E., 1983, Mitoxantrone: a review of experimental and early clinical studies, *Cancer Treatment Reviews*, 10:103.

Wertz, J.E., 1972, "Electron Spin Resonance: Elementary Theory and Practical Applications," McGraw-Hill, New York.

OXIDATIVE EFFECTS OF PHYSICAL EXERCISE

Alexandre T. Quintanilha

Applied Science Division
Lawrence Berkeley Laboratory, and
Physiology-Anatomy Department
University of California
Berkeley, California 94720, U.S.A.

INTRODUCTION

Increased physical activity in animals is accompanied by significantly higher rates of oxygen consumption (see Brooks & Fahey, 1984); in man these increases can be of the order of ten-fold. While the biological toxicity of oxygen was already known to Lavoisier and Priestley and has been clearly demonstrated for animals at higher pressures (> 1 atmosphere) and at higher concentrations ($> 20\%$), less attention has been paid to the effect of higher rates of consumption. For flying insects (Sohal, 1981) several interesting studies have clearly established a positive correlation between higher rates of oxygen consumption (due to increased physical activity) and a shorter life span (accompanied by a faster rate of accumulation of oxidative products of lipids and proteins). In mammals such a correlation has not been established, perhaps for the simple reason that the increase in the rate of oxygen consumption of mammals at the onset of exercise is significantly smaller than in flying insects.

We have attempted to understand the oxidative consequences to a number of different tissues of increased levels of oxygen consumption due to physical activity. Most of our work has been done on rats, but we believe that our results could be extremely useful in trying to understand what happens in humans. Our findings indicate that there is significant oxidative damage to both skeletal muscle and liver during physical exercise, and that the liver may play an important role in supplying the muscle with the required glutathione for its increased anti-oxidative needs during exercise.

Evidence for Tissue Damage During Physical Exercise

The literature contains many examples of tissue damage that results from increased levels of physical activity in a variety of different animal species. Considering that exercise can lead to temperature increases in many tissues (as high as 5 to 6 degrees centigrade in muscle in man), dehydration and hormonal alterations, it is not surprising that many tissue functions are significantly altered. A single bout of exhaustive exercise causes focal necrosis, inflammation and increased lysosomal acid hydrolase activity in skeletal muscle (Vihko et al, 1978 & 1979); it can also lead to increased mechanical fragility of rat liver lysosomes, and protein degradation mediated via the release of a large number of different lysosomal enzymes (Kasperek et al, 1980 &1982).

Exercise-induced injury has also been shown to result in increased serum levels of the intracellular enzyme creatine kinase (Armstrong et al, 1983; Newman et al, 1983a), structural cellular changes observed with both light and electron microscopy (Friden et al, 1983; Highman & Altland, 1963), and decreases in the ability to develop force, that cannot be simply attributed to fatigue (Davies & White, 1981; Newman et al, 1983a). However, the most familiar symptom of injury that is associated with exercise is the phenomenon of delayed muscle soreness which lasts for 1-5 days after strenuous or unaccustomed exercise (Asmussen, 1956; Hough, 1902; Newman, 1983b). Several causes have been invoked as responsible for exercise-induced injury to skeletal muscle; these include large increases in muscle hydrogen ion and lactate concentrations, and mechanical disruption of muscle fibers. Lines of evidence supporting each of these hypotheses are reviewed in a recent paper by McCully (1986). The purpose of this chapter is to review the evidence that we have accumulated in the last couple of years that supports a different mechanism of exercise-mediated tissue (including muscle) injury, namely oxidative injury.

INITIAL STUDIES

In 1982 (Davies et al, 1982) we reported that extensive exercise results in decreased mitochondrial respiratory control, loss of sarcoplasmic reticulm (SR) and endoplasmic reticulum (ER) integrity, and increased levels of lipid peroxidation products in both the liver and skeletal muscle of rats (see Table 1). We also observed a two- to three-fold increase in free radical (R·) concentrations in both tissues following exercise to exhaustion. While at the time we had very little idea of what R· might be, subsequent work showed that it might very well be due to the ubiquinone radical from the mitochondrial inner membrane. Such a finding is not surprising since the ubiquinone radical has been shown to be reasonably stable (particularly when the molecule is bound to a protein) and quinones are known to be good generators of superoxide radicals when undergoing redox cycling.

Similar measurements done on Vitamin E deficient rats (Table 2) showed the same general pattern of results except that the damage was already more pronounced for the resting animals, and the endurance capacity of these animals was reduced by about 40% when compared to (Vitamin E sufficient) control animals.

In a related study (Aikawa et al, 1984), we investigate the depletion of Vitamin E from muscle and liver of four groups of rats: two of these groups were on a Vitamin E deficient and the other two were on a Vitamin E sufficient (control) diet. Of the two groups on each of the diets, one was exercised daily (endurance trained) over a period of nine weeks (increasing the running time each day until the animals ran for about two hours a day); the other group was not exercised (sedentary).

Table 1. Effects of exercise to exhaustion on liver and skeletal muscle tissue homogenates [data taken from Table 1 of Davies et al., 1982].

Parameter Measured	Muscle		Liver	
	Rested	Exercised	Rested	Exercised
Pyruvate-Malate RCI	4.2	3.2	3.5	3.0
Glutamate RCI	3.8	3.0	4.0	3.3
Succinate RCI	1.2	1.0	6.2	3.9
SR or ER % Latency	53.5	20.9	52.6	38.4
Lipid Peroxidation	27.7	50.1	32.9	77.3
R· Concentration	8.0	17.0	8.3	19.5

Respiratory control indices (RCI) were calculated as the rate of uncoupled respiration/rate of basal respiration. Sarcoplasmic reticulum (SR) and encoplasmic reticulum (ER) percent latencies are expressed as [(total solubilized activity - initial activity)/(total solubilized activity)] $\times$ 100. Lipid peroxidation is reported as nanomols malondialdehyde/g wet tissue. Free radical (R·) concentrations are electron spin resonance signal peak heights at g $\sim$ 2.004 in arbitrary units.

Table 2. Effects of exercise to exhaustion on liver and skeletal muscle tissue homogenates in Vitamin E defficient animals [data taken from Table 1 of Davies et al., 1982].

Parameter Measured	Muscle		Liver	
	Rested	Exercised	Rested	Exercised
Pyruvate-Malate RCI	3.5	2.9	3.3	2.7
Glutamate RCI	3.3	2.3	3.7	2.3
Succinate RCI	1.1	1.0	2.8	2.2
SR or ER % Latency	37.9	11.8	48.2	14.5
Lipid Peroxidation	39.3	45.1	61.0	79.8
R· Concentration	10.6	13.7	11.9	14.7

Parameters are the same as in Table 1.

189

Table 3. Tissue levels of Vitamin E (μg/g wet tissue) in homogenates of liver and total hindleg muscle of rats fed a Vitamin E sufficient or a Vitamin E deficient diet [data taken from Table 1 of Aikawa et al., 1984].

	Trained	Sedentary
Liver		
Control Diet	37.79	34.55
E-deficient Diet	0.15	2.96
Muscle		
Control Diet	4.91	7.20
E-deficient Diet	0.00	1.17

We were able to show (Table 3) that muscle tissue levels of Vitamin E were significantly lower in endurance-trained rats than in sedentary animals, whether these were fed a Vitamin E deficient or control diet. In Vitamin E deficient rats, liver tissue levels of Vitamin E were significantly lower in those that were endurance-trained than in those that were sedentary; this was not the case in control animals.

These results indicate that animals undergoing an exercise program probably have a higher requirement for Vitamin E than sedentary animals, and that while such additional requirements are undisputable for animals on a Vitamin E deficient diet, there are additional requirements also observed in animals on a "supposedly" Vitamin E sufficient (normal) diet.

These studies provided strong evidence of oxidative damage to skeletal muscle and liver during bouts of physical exercise. As pointed out in the chapter by Drs. Jackson and Edwards in this volume (pp. 197-210), however, we still do not know how such increases in lipid peroxidation and depletion of Vitamin E in these tissues are related to the observed increases in free radical concentrations.

GLUTATHIONE STUDIES

Glutathione plays a major role in many cellular and organismal functions: among these it provides cellular protection against oxidative damage, is an important intermediate in pathways of drug detoxification and amino acid transport, modulates enzymatic activity (particularly of those enzymes with critical SH residues), and plays a role in Ca^{2+} homeostasis (for reviews see: Larsson et al, 1983; Sies & Wendel, 1978; Meister & Anderson, 1983). Conditions that perturb intracellular levels of glutathione have been shown to result in significant alterations in cellular metabolism (see also chapter by Dr. Smith in this volume). Reduced glutathione (GSH) is usually present intracellularly in the millimolar range while the concentration of its oxidized form, glutathione disulfide (GSSG), is two or three orders of magnitude lower (Kosower & Kosower, 1978). Increased oxidative stress to tissues usually results in decreased levels of intracelluar GSH and increased GSSG. Since GSSG is either exported from the cells or reduced back to GSH (via the NADPH requiring enzyme glutathione reductase), its intracellular levels usually remain well below GSH levels, even under conditions of severe oxidative stress to the cells.

We have investigated the effect of physical exercise on the ability of cells to protect themselves against oxidant stress by the resulting changes in the intracellular ratios of GSH to GSSG (Lew et al, 1985; Pyke et al, 1986).

Characteristic values for GSSG and total glutathione (GSH + GSSG) in the plasma, liver and skeletal muscle of rested control and exercised rats are shown in Table 4. All exercised animals in these studies were run on a Quinton rodent treadmill at a grade of 8.5% (15%) and at a speed of approximately 24.1 m/min (0.9 mph) for 90-120 minutes. The control values of GSSG and total glutathione that we report are in good agreement with published results; day to day variations reported in our studies (Lew et al, 1985) are probably due to the fact that our experiments were not always performed at the same time of day and circadian rhythms in the concentration of glutathione are well known, at least in the liver.

Table 4 shows that as we had expected, physical exercise results in higher levels of GSSG and lower levels of GSH in the liver or skeletal muscle in rats, when compared to sedentary controls. In the plasma, the levels of total glutathione are always low (3-6 μM) for resting animals, but increase with the onset of exercise; such increases are not only due to large increases in levels of GSSG, but in GSH as well.

While the rise in plasma GSSG is not surprising in view of the well-known mechanism of GSSG export out of cells when intracellular levels rise, the increase in plasma GSH with exercise is somewhat more puzzling, but perhaps more interesting.

Whereas we cannot exclude the possibility that the rise in plasma GSH is due to cell damage (particularly red blood cells which are damaged during exercise), we feel that such a rise may be in part the result of an exercise-induced stimulation of GSH efflux

Table 4. Tissue levels of glutathione disulfide (GSSG) and total glutathione (GSH + GSSG = GS) [data taken from Table 1 of Lew et al., 1985].

	PLASMA*			LIVER**			MUSCLE**		
	Cont.	Exer.	% Change	Cont.	Exer.	% Change	Cont.	Exer.	% Change
GSSG	0.13	0.47	+251	0.007	0.014	+103	0.0006	0.0012	+100
GS	3.56	4.60	+29	3.15	2.58	-18	0.33	0.31	-6

Cont. = Control sedentary values.
Exer. = Values after a bout of exercise.
* Values expressed in μmol/liter.
** Values expressed in μmol/g wet tissue.

Table 5. Ratios of glutathione disulfide (GSSG) to total glutathione (GSSG + GSH) [data taken from Table 2 of Lew et al., 1985].

PLASMA			LIVER			MUSCLE		
Cont.	Exer.	% Change	Cont.	Exer.	% Change	Cont.	Exer.	% Change
0.0374	0.1020	+173	0.0022	0.0054	+145	0.0018	0.0039	+117

Cont. = Control sedentary values.
Exer. = Values after a bout of exercise.

from the liver. Independent studies (Sies & Graf, 1985) have shown that GSH efflux across the sinusoidal plasma membrane in an isolated perfused rat liver preparation can be induced by increasing vasopressin levels in the perfusing fluid (within physiological limits). The well-known increases in plasma vasopressin levels during physical exercise would then provide the mechanism for export of GSH from the liver into the plasma. It could then be transported to the muscles, where it is mostly needed to protect against the increased oxidative stress that results from increased muscular activity. Were this to be the case, the liver could play a more important role in physical exercise than is generally appreciated.

As can be seen from Table 5, physical exercise increases the ratio of glutathione disulfide to total glutathione in all three tissues; were this to be true in general, it could provide a simple and rapid way to assess whole body oxidative stress in humans by measuring glutathione (reduced and oxidized) levels in the plasma. As will be shown later, however, the changes in plasma levels of glutathione in response to exercise-induced oxidant stress depends on many factors, namely, the duration and intensity of exercise, body weight, age, and level of physical fitness.

Subsequent work (Pyke et al, 1986) has shown that during extensive physical exercise skeletal muscle is able to maintain relatively high levels of total glutathione (always > 60% of control), while the liver is depleted to as much as 20% of control values. Also the initial increase in total glutathione in the plasma, observed at the onset of exercise, is followed by a progressive decline (to levels that remain above 60% of control) as the bout of exercise is prolonged.

In addition, we have found that the depletion of total glutathione in the liver persists and is further exacerbated for several hours following cessation of physical exercise (Lew et al, 1987). Since oxygen consumption remains high following a bout of exercise, and if reduced glutathione is oxidized via a mechanism associated with higher rates of oxygen consumption (and perhaps, therefore, higher rates of oxygen radical generation), our findings are not surprising.

The observation that prolonged exercise results in a severe depletion in liver glutathione while glutathione levels are kept relatively high in skeletal muscle, provides additional evidence, albeit indirect, for our proposal that one of the important functions of the liver is to provide glutathione to muscle during exercise.

EFFECTS OF ENDURANCE TRAINING ON GLUTATHIONE RESPONSE TO EXERCISE-INDUCED OXIDANT STRESS

Results in Table 6 show that following a short running bout (42 min.), endurance-trained rats, in contrast to the results that we had obtained previously for untrained animals, show no significant changes in tissue levels of GSSG for plasma, liver, heart and skeletal muscle; in many cases, e.g., for plasma, liver and heart, there is a decrease in the levels of GSSG following this short bout of exercise in endurance-trained animals.

Furthermore, while significant increases in plasma total glutathione have been demonstrated for untrained animals following short running bouts (< 100 min.) (Lew et al, 1985), the same is not true for endurance-trained animals (Table 6).

It appears therefore, that at least during the initial stages of exercise, an endurance-trained rat shows none of the evidence of increased oxidative stress that we have so clearly demonstrated for untrained animals undergoing comparable work loads. Whether this is due to the induction of other more efficient mechanisms of defense against oxidative stress in the trained animals, or whether it is simply the result of faster

Table 6. Effect of a 42 minute running bout on the levels of total glutathione (GSH + GSSG) and glutathione disulfide (GSSG) of plasma, liver, heart and skeletal muscle in endurance trained and untrained rats. Number of animals used: 7 untrained resting controls, 7 untrained runners, 5 trained resting controls, and 6 trained runners [results taken from Lew et al., 1987].

TISSUE TRAINING STATE	GLUTATHIONE LEVELS (GSH + GSSG)		(GSSG)	
	Rested	42 min run	Rested	42 min run
Plasma*				
Untrained	8.80	20.52	0.54	3.84
Trained	6.24	5.76	0.66	0.36
Liver**				
Untrained	3.04	3.14	.0027	.0079
Trained	2.80	2.61	.0036	.0026
Heart**				
Untrained	0.65	0.58	.0033	.0041
Trained	0.71	0.66	.0019	.0010
Skeletal Muscle**				
Untrained	0.39	0.29	.0036	.0054
Trained	0.41	0.35	.0023	.0031

* Values expressed in μmol/liter.
** Values expressed in μmol/g wet tissue.

turnover of glutathione in the trained aimals (that could also be regarded as a more efficient mechanism of defense against oxidative stress) is unclear at this stage. Some evidence suggests that many enzymes can be induced in liver and muscle cells following a program of endurance-training in rats; the literature, however, is confusing in this regard.

CONCLUSION

There is no doubt that strenuous or unaccustomed exercise can cause injury not only to skeletal muscle, but also to liver and perhaps several other organs.

The evidence that we have gathered thus far strongly suggests that such damage might be partially due to increased oxidant stress to these tissues. Whether such effects are the result of well-known free radical mechanisms of damage to biological materials such as lipids, proteins and DNA is not clear at present.

The body, however, has a variety of mechanisms at its disposal to increase its protection against oxidant stress to different tissues. Endurance training, for instance, may increase the efficiency of antioxidative pathways in cells.

Perhaps more interesting though is the fact that during extensive exercise, the liver, in addition to supplying glucose to muscle, may also supply glutathione as an protective antioxidant. After all, it seems that the role of the liver supplying glutathione to other tissues (e.g., lung, kidney, etc.) has been fairly well documented (see Kaplowitz et al., 1985). These results may be particularly relevant when attempting to understand the effects of physical exercise on a large number of cellular and organismal functions that are well-known to depend critically on the glutathione status of the liver. It is this particular area of study that we intend to pursue.

REFERENCES

Aikawa, K.M., Quintanilha, A.T., deLumen, B.O., Brooks, G.A., and Packer, L., 1984, *Bioscience Reports*, 4:253.

Armstrong, R.B., Laughlin, M.H., Rowe, L., and Taylor, C.R., 1983, *J. Appl. Physiol.*, 55:518.

Asmussen, E., 1956, *Acta Rheumatol. Scand.*, 2:109.

Brooks, G.A., and Fahey, T.D., 1984, "Exercise Physiology: Human Bioenergetics and Its Applications," John Wiley & Sons, New York.

Davies, C.T.M., and White, M.J., 1981, *Pfluegers Arch.*, 392:168.

Davies, K.J.A., Quintanilha, A.T., Brooks, G.A., and Packer, L., 1982, *Biochem. Biophys. Res. Commun.*, 107:1198.

Friden, J., Sjöstrom, M., and Ekblom, B., 1983, *Int. J. Sports Med.*, 4:170.

Highman, B., and Altland, P.D., 1963, *Am. J. Physiol.*, 205:162.

Hough, T., 1902, *Am. J. Physiol.*, 7:76.

Kaplowitz, N., Aw, T.Y., and Ookhtens, M., 1985, *Ann. Rev. Pharmacol.*, 25:715.

Kasperek, G.J., Dohm, G.L., Barakat, H.A., Strausbach, P.H., Barnes, O.W., and Snider, R.D., 1982, *Biochem. J.*, 202:281.

Kasperek, G.J., Dohm, G.L., Tapscott, E.P., and Powell, T., 1980, *Proc. Soc. Exp. Biol. Med.*, 164:430.

Kossower, N.S., and Kossower, E.M., 1978, *Int. Rev. Cytol.*, 54:109.

Larsson, A., Orrenius, S., Holmgren, A., and Mannervik, B., eds., 1978, "Functions of GSH," Raven Press, New York.

Lew, H., Pyke, S., and Quintanilha, A., 1987, *J. Bioelectrochem.*, (in press).

Lew, H., Pyke, S., and Quintanilha, A., 1985, *FEBS Lett.*, 185:262.

McCully, K.K., 1986, *Fed. Proc.*, 45:2933.

Meister, A., and Anderson, M.E., 1983, *Ann. Rev. Biochem.*, 52:711.

Newman, D.J., Jones, D.A., and Edwards, R.H.T., 1983a, *Muscle Nerve*, 6:380.

Newman, D.J., Mills, K.R., Quigley, B.M., and Edwards, R.H.T., 1983b, *Clin. Sci.*, 64:55.

Pyke, S., Lew, H., and Quintanilha, A., 1986, *Biochem. Biophys. Res. Commun.*, 139:926.

Sies, H., and Graf, P., 1985, *Biochem. J.*, 226:545.

Sies, H., and Wendel, A., eds., 1978, "Functions of GSH in Liver and Kidney," Springer-Verlag, New York.

Sohal, R.S., 1981, "Age Pigments," Elsevier, Amsterdam.

Vihko, V., Rantamäki, and Salminen, A., 1979, *J. Appl. Physiol. Respirat. Environ. Exercise Physiol.*, 47:43.

Vihko, V., Rantamäki, and Salminen, A., 1978, *Histochemistry*, 57:237.

FREE RADICALS, MUSCLE DAMAGE
AND MUSCULAR DYSTROPHY

M.J. Jackson and R.H.T. Edwards

Department of Medicine
The University of Liverpool
Royal Liverpool Hospital
Liverpool, L69 3BX

INTRODUCTION

The genetically inherited muscular dystrophies are chronic degenerative diseases of muscle which, in the more severe form (Duchenne), usually result in death in the second or third decade of life. All attempts to modify therapeutically the expression of the defective gene in patients have been limited by a lack of knowledge of the nature of the basic biochemical defect in these disorders.

It has been recognised for a number of years that the severest forms of the muscular dystrophies share many characteristics with vitamin E or selenium deficiency myopathy of animals, although attempts to treat patients with vitamin E have generally been unsuccessful (Fitzgerald & McArdle, 1941; Walton & Natrass, 1954). The recognition that at least one of the major functions of vitamin E and selenium within the body is to inhibit the toxic effects of free radical mediated processes has prompted a reinvestigation of this area.

In addition, we have also been studying an alternative approach by which therapies for these disorders might be devised. This is to try to reduce the amount of damage to the muscle fibres which results from the basic biochemical defect. Patients with muscular dystrophy have histological changes in their muscle indicative of an increased rate of loss of muscle fibres and have plasma activities of muscle - derived enzymes [e.g., creatine kinase (CK)] up to 3 orders of magnitude greater than normal. However, any attempt to interfere with the process of muscle cell damage requires a knowledge of the biochemical mechanisms underlying this process. To this end we have been studying these mechanisms using both *in vitro* and *in vivo* preparations. These studies have suggested that free radical processes may play some role in the more fundamental general process of muscle damage and hence may be relevant to muscular dystrophy, whatever the fundamental underlying defect in this disorder.

This chapter will review studies investigating the mechanisms of damage to skeletal muscle, the techniques which have been used to examine free radical mediated reactions in muscle tissue and will conclude with a discussion of the work which has been undertaken to examine antioxidant metabolism and free radical mediated processes in dystrophic tissue, and the implications of these studies for therapy in patients with these disorders.

STUDIES OF THE BIOCHEMICAL MECHANISMS UNDERLYING SKELETAL MUSCLE DAMAGE

Model Systems for the Study of Muscle Damage

Despite extensive use of isolated heart preparations to study damage to cardiac muscle, there have been relatively few studies using skeletal muscle. Those studies which have been undertaken have primarily relied on one indicator of the extent of damage, i.e., either histological and electron microscopic appearance (Duncan, 1978; Publicover et al., 1978) changes in the rate of protein breakdown (e.g., Rodemann et al., 1981) or variations in the rate of efflux of cytoplasmic enzymes (Dawson, 1966; Suarez-Kurtz & Eastwood, 1981). We have used a system which utilises small mouse extensor digitorum longus muscles from which the release of intracellular components can be measured as an index of damage (Jones et al., 1983); muscles from this sytem can also be taken for examination of their electron microscopic appearance (Jones et al., 1984). We have also examined damage to muscle *in vivo*, by monitoring blood creatine kinase activities following excessive contractile activity in specific muscles (Jackson et al., 1983a).

These studies have indicated that two different types of biochemical processes are associated with damage to skeletal muscle. These are calcium-mediated degradative mechanisms and free radical-mediated oxidation processes.

Calcium-Mediated Damage to Myofibrils

Changes in the intracellular calcium content have been implicated in the mechanisms by which damage occurs to several tissues, including skeletal muscle. In cardiac tissue, damage due to hypoxia or reoxygenation has been shown to be associated with an increase in tissue calcium content (Nayler et al., 1979), while in hepatocytes loss of cell viability following incubation with various toxins has been reported to be dramatically reduced when the external calcium is removed from the incubation fluid (Schanne et al., 1979); this finding has subsequently been the subject of much controversy (e.g., Smith et al., 1981; Fariss et al., 1985).

High external calcium concentrations (3-10 mmols per litre) have been shown to increase CK release from resting animal (Soybell et al., 1978) and human skeletal muscle (Anand & Emery, 1980). Treatment of skeletal muscle preparations with the calcium ionophore A23187 has further demonstrated the potential of increased calcium levels to induce damage (Duncan et al., 1979). Slow calcium channel blocking agents (i.e., calcium antagonists) have been found to reduce the CK release from human skeletal muscle *in vitro* (Anand & Emery, 1982) and other workers have shown that alternative manipulations designed to reduce calcium accumulation in skeletal muscle (i.e., parathyroidectomy) prevents the pathological changes to skeletal muscle in hampsters with congenital forms of muscular dystrophy (Palmieri et al., 1981). In another pathological condition of muscle (selenium deficiency myopathy) ^{45}Ca accumulation by muscles have been reported to precede biochemical, histological, or clinical evidence of myopathy (Godwin et al., 1975).

We have used an *in vitro* system to examine the role of external calcium in the release of enzymes from skeletal muscle (Jones et al., 1983). It was found that release of enzymes following different stresses (e.g., excessive contractile activity, treatment with low dose detergents, or treatment with mitochondrial inhibitors) could be prevented by removal of the external calcium during the damaging procedure (Jones et al., 1984; Jackson et al., 1984a). It was also found that this manipulation was equally effective in protection of muscle against the histological and electron microscopic changes induced by excess contractile activity (Jones et al., 1984). Other experiments with this sytem have also demonstrated a dramatic increase in total muscle calcium during either excess contractile activity leading to enzyme efflux or treatment with mitochondrial inhibitors (e.g., dinitrophenol), (Claremont et al., 1984).

These results suggest that damage to skeletal muscle is accompanied by an influx of extracellular calcium down the large extracellular to intracellular concentration gradient for this element. This increased intracellular calcium content then mediates further pathological changes. Despite the fact that others have claimed an effect of calcium antagonists in skeletal muscle (Anand & Emery, 1982), we have been unable to demonstrate any protective effect of these agents in isolated preparations (Jones et al., 1984).

Considerable speculation has surrounded the possible mechanisms by which increased calcium may mediate pathological processes in cells. The hypotheses which have been proposed regarding skeletal muscle include accumulation of calcium by mitochondria leading to loss of oxidative energy production (Wrogmann & Pena, 1976), activation of calcium dependent proteases (Ebashi & Sugita, 1979), activation of lysosomal proteases by the stimulation of prostaglandin production (Rodemann et al., 1981) or direct release of lysosomal enzymes (Duncan, 1978).

Inhibitor studies which we have performed suggest that calcium influx may be a key step in the damage resulting from the activation of phospholipase A (Jackson et al., 1984a). Activation of this enzyme will result in the breakdown of membrane phospholipids leading to production of lysophospholipids and free fatty acids. Accumulation of lysophospholipids will lead to a breakdown of membrane lipid organizaton (Weglicki, 1980) and the free fatty acids released will act as detergents, causing membrane damage (Katz, 1982). In addition, among the free fatty acids produced will be arachidonic acid. This is the precursor of the prostaglandin series of compounds. Prostaglandins have been reported to be involved in the control of muscle protein homeostasis (Rodemann et al., 1981; Baracos et al., 1983), and an excess production of these may further compromise muscle integrity.

Several workers have implicated an increase in free radical-mediated reactions in the damage to muscle which accompanies exercise. Brady and co-workers (1979) studied the response of rats to exhaustive swimming exercise and found that both liver and muscle tissue contained increased amounts of malonaldehyde (a product of free radical-mediated lipid peroxidation) following exercise and suggested an increased amount of free radical intermediates may be produced during exercise. Tappel's group has studied pentane excretion in the breath of normal subjects (Dillard et al., 1978) and rats (Gee & Tappel, 1981) during exercise (pentane is a product of the free radical-mediated peroxidation of certain fatty acids) and has found a large increase, suggesting increased lippid peroxidation at some site in the body during exercise. Dillard et al. (1978) have also demonstrated an apparent protective effect of supplemental vitamin E against this process. The major role of vitamin E in the body appears to be to act as a lipid soluble antioxidant preventing free radical-mediated peroxidation of membrane components. Packer and co-workers have been specifically examining the effects of exercise in vitamin E deficient animals (Quintanilha & Packer, 1983), and they claim that these animals have a considerably lower tolerance of exercise since they are less able to withstand the oxidative stress which occurs with increased mitochondrial energy metabolism. These workers have also demonstrated increased levels of lipid peroxidation products (malonaldehyde) in animal tissues following exercise (Davies et al., 1982).

It has been pointed out that damage to tissues can be the cause of lipid peroxidation as well as the consequence of it (Halliwell & Gutteridge, 1984), and the studies examining lipid peroxidation products following exercise may well fall into this category. An alternative approach is to try and examine free radical intermediates directly in tissues using physical methods. Davies et al. (1982) have shown that strenuous exercise leads to an increased electron spin resonance signal from muscle and liver tissue. We have also examined electron spin resonance signals from skeletal muscle during experimental skeletal muscle damage induced by excessive contractile activity *in vivo* (Jackson et al., 1985a). In these studies, a $70 \pm 20\%$ increase in the stable electron spin resonance signal was shown to be associated with an increase in the plasma CK activity of the rat following exercise. Unfortunately, in this situation it is still not possible to say which of the increased free radical concentration or the damage to the muscle is primary or whether the two findings are merely coincidental.

We have also used the *in vitro* skeletal muscle damage system (Jones et al., 1983) to examine this area. It has been demonstrated that the vitamin E content of the muscle influences the amount of lactate dehydrogenase enzyme released from muscles following an equivalent amount of contractile activity both *in vitro* and *in vivo* (Jackson et al., 1983a).

The hypothesis that damage to skeletal muscle is associated with increased free radical activity which can be prevented by vitamin E also receives support from work with cardiac muscle. Guarnieri et al. (1978) have shown that supplemental vitamin E protects the heart muscle against reperfusion induced injury, while increased free radicals have been extensively implicated in damage due to hypoxia and/or reperfusion injury (McCord, 1985; Rao, et al., 1983). In addition, increased free radical-mediated damage has been implicated in the deleterious consequences of various toxic agents on other tissues (Halliwell & Gutteridge, 1985).

Possible Integrating Hypotheses

It is possible that the two previously proposed mechanisms are independently activated during damage, but they may be linked by one of several different pathways; McCord (1985) has suggested that calcium accumulation (in ischaemic tissue) promotes the conversion of the enzyme xanthine dehydrogenase to xanthine oxidase which then acts upon hypoxanthine formed from the breakdown of adenosine metabolites to produce uric acid and superoxide radicals. Superoxide can then cause damage to tissues, either directly or via production of highly reactive hydroxyl radicals. However, xanthine oxidase has so far not been detected in skeletal muscle, and this is therefore unlikely to be the mechanism by which damage occurs in skeletal muscle. An alternative mechanism concerns the action of calcium on mitochondria, since generation of hydrogen peroxide was found to be enhanced in antimycin A supplemented rat heart mitochondria in the presence of increased calcium levels (Cadenas & Boveris, 1980). This production of hydrogen peroxide may occur from the superoxide dismutase catalysed conversion of superoxide radicals generated by the ubiquinone of the mitochondrial membrane (Boveris & Chance, 1973); Boveris et al., 1976). In addition, various workers have demonstrated a direct effect of calcium upon the susceptibility of membrane lipids to peroxidation (Jain & Shohet, 1981; Gutteridge, 1977).

We have proposed an alternative pathway whereby calcium accumulation may produce increased free radical damage in muscles. The free fatty acid released from membranes by calcium activation of phospholipase A is likely to be more susceptible to free radical-mediated lipid peroxidation than membrane-bound polyunsaturated fatty acids. Membrane-bound fatty acids are intimately associated with vitamin E, whereas once released there are no similar cytoplasmic lipid-soluble antioxidants available to protect the molecule. However, the cytoplasm may well be the major site of free radical production since several enzyme systems produce reactive oxygen metabolites as a normal component of metabolism. Peroxidation of the free polyunsaturated fatty acids may therefore result (Jackson et al., 1985b).

DETECTION AND IDENTIFICATION OF FREE RADICALS IN MUSCLE TISSUE

In order to demonstrate that increased free radical activity plays some role in the processes underlying experimental skeletal muscle damage and/or the pathogenesis of muscle breakdown seen in neuro-muscular diseases such as Duchenne muscular dystrophy, it is necessary to be able to provide evidence of their increased activity in the appropriate muscle samples. This is a technically difficult procedure since there is still no single accepted test for evidence of increased free radical activity in biological material. At the present time, it is generally accepted that a pathological process can be ascribed to a free radical-mediated mechanism if there is evidence of increased free radical activity within the material under study; if it can be prevented by the use of specific inhibitors of free radical activity (i.e., antioxidants or protective enzymes); and if the process can be mimicked by an exogenous source of free radicals. Damage to skeletal muscle can undoubtedly occur if muscle is stressed with a free radical generating system (Jackson et al., 1983b) and the methods we have used to try and demonstrate a pathogenic role of free radicals in experimental muscle damage are:

(1) Examination of model systems for a protective effect of antioxidants.

(2) Measurement of indirect indices of free radical activity.

(3) Direct detection of free radicals by physical methods.

Protective Effect of Antioxidants

The effect of a variety of antioxidants on experimental skeletal muscle damage is shown in Table 1. These agents have all been claimed to have the ability to inhibit free radical-mediated processes, but only a certain number of these were able to prevent the damage to skeletal muscles. We originally utilised chlorpromazine, mepacrine and dibucaine because of their ability to inhibit phospholipase enzymes (Jackson et al., 1984a), but later found they had significant alternative properties such as free radical scavenging (Jackson et al., 1984b).

Table 1. Effect of antioxidants on skeletal muscle damage.

Protective	No Effect	Damaging
Vitamin E	Vitamin C	Menadione
Chlorpromazine	Butylated hydroxytoluene	
Mepacrine	Butylated hydroxyanisole	
Dibucaine	Propylgallate	
	Zinc	
	(Allopurinol)	

There are a variety of different reasons why one antioxidant may be more efficient at inhibition of a certain free radical-mediated process than others. It is necessary for the inhibitor to reach an appropriate site at a sufficiently high concentration to prevent the toxic reactions of a sufficiently large number of radical species. Therefore, differences in lipid solubility, sight of subcellular concentration, and ability to inhibit (scavenge) different radicals may explain the lack of effect of those agents shown in Table 1 compared to those which are effective.

These studies serve to illustrate the difficulties of allocating a crucial role for free radicals in any complex process on the basis of a protective effect of antioxidants. Because of differing solubilities and other properties, it is unlikely that all antioxidants will inhibit a particular process and it must be recognised that most substances have a variety of actions in intact cells. In many circumstances a protective effect of complex antioxidants can therefore only be thought of as supportive evidence of the involvement of free radicals.

Measurement of Indirect Indices of Free Radical Activity

By the very nature of their toxic effect, free radicals interact rapidly with a variety of cellular components. In certain circumstances the nature of the products of these reactions are well described. The interaction of various different radical species with polyunsaturated lipids in the process of lipid peroxidation is one such relatively well understood system, and the products or intermediates of this reaction are those most frequently monitored as indirect indices of free radical activity.

A simplified schematic representation of the process of free radical-mediated lipid peroxidation is presented in Figure 1, which also shows the substances which can be monitored in order to study this process. In practice, certain measurements such as quantification of lipid peroxides are rarely undertaken because of the insensitivity and difficulty of these analyses. The relative benefits of the other parameters are discussed elsewhere in this book (see chapter by Dr. Slater).

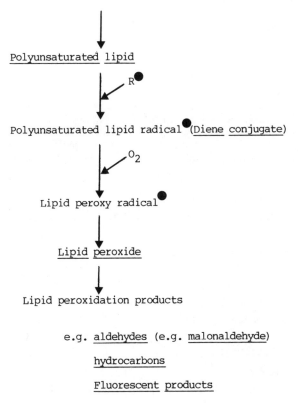

Figure 1. Simplified schematic representation of the process of lipid peroxidation. The substances which can be measured in order to monitor lipid peroxidation are underlined.

Our studies designed to examine indices of lipid peroxidation in the experimentally damaged skeletal muscles have revealed a transient rise in the malonaldehyde content of mouse soleus muscle immediately following damaging contractile activity (Table 2), but that this is not maintained and one hour after the end of stimulation levels are not significantly different to incubated, nonstimulated muscles.

Dystrophic muscle from human subjects obtained using the needle biopsy technique has also been found to contain increased amounts of malonaldehyde (Table 3). This finding was originally reported by Kar & Pearson in 1979 and was later confirmed by Jackson et al. (1984c).

Table 2. Mouse soleus malonaldehyde content following damaging contractile activity.

	0 mins post-stimulation	*60 mins post-stimulation*
Stimulated Muscle	62.6 ± 6.3	43.6 ± 2.4
Control, Non-stimulated Muscles	33.0 ± 2.5	39.8 ± 4.3

All results are expressed as mean±SEM of 6 muscles and presented as nmoles/g wet wt.

Stimulated muscles were incubated for 30 mins prior to 30 mins stimulation with square wave pulses of 0.1 msec duration and 30V; muscles were stimulated for 0.5 sec every 5 sec at a frequency of 100 Hz. Fresh mouse soleus muscle was found to have a malonaldehyde content of 32.2±7.2 nmole/g wet wt. Malonaldehyde was measured using the thiobarbituric acid test.

Table 3. Malonaldehyde (or TBA reactive products) in human dystrophic muscle.

	Normal Subject	*Dystrophic*
Data from Jackson et al, (1984) (nmoles malonaldehyde /g protein)	393.0 ± 58.7	1340.6 ± 312.4
Data from Kar & Pearson (1979) (A_{532}/g wet wt.)	0.70 ± 0.10	1.67 ± 0.20

Unfortunately, malonaldehyde is an unreliable index of lipid peroxidation in intact biological tissues since it is an intermediate rather than an end product of lipid peroxidation, having been shown to be metabolised by intact cells (Siu & Draper, 1982). Normal levels of malonaldehyde are therefore inconclusive. Elevated levels are somewhat more reliable, although they can only be used as a qualitative guide rather than an accurate indication of the amount of lipid peroxidation which has occurred.

Other workers have studied the production of volatile hydrocarbons following exercise in animals and man in studies which may be of relevance to those of experimental muscle damage discussed here. Gee & Tappel (1981) examined pentane excretion in animals subjected to exhaustive exercise, while Dillard et al. (1978) studied the effects of exercise and vitamin E on pentane excretion in man. Both of these workers reported increased pentane excretion during exercise, suggesting an increased rate of lipid peroxidation in the exercising species. However, neither workers appear to have considered the possible effects of the increased blood flow to tissues which occurs during exercise. If volatile hydrocarbons are retained in relatively poorly perfused tissues at rest, then the increased blood flow would be likely to induce release of stored pentane from these sources. Snider et al. (in press) have recently studied isopentane (a foreign gas with no known biological origin) excretion during exercise and have found it to be increased by a similar extent to n-pentane and suggested that release of exogenously stored gas was the likely source of the increase in both isomers, rather than metabolic production.

One further problem bedevils attempts to ascribe a degenerative process to an increased free radical production. Halliwell & Gutteridge (1984) have claimed that damaged tissues can induce increased free radical production and hence the production of lipid peroxidation products rather than the reverse situation.

Direct Detection of Free Radicals by Physical Methods

Direct detection of free radicals can be achieved by electron spin resonance spectrometry (ESR). This technique has had wide applications in the field of physical chemistry, but has only been relatively sparsely used to examine biological materials. Its use in the study of biological processes has been limited by the high water content of biological materials since water strongly absorbs microwaves. Early investigators attempting to utilise this technique in the study of tissues examined lyophilised material (Commoner et al., 1954). It later became apparent that signals obtained from dried material frequently resulted from the lyophilisation process or from subsequent changes (Heckley, 1976). Some studies have been undertaken on fresh tissue (e.g., Davies et al., 1982), but a more satisfactory solution to the problem seems to be rapid freezing of the tissue; this should provide esr spectra of the radicals as they were at the time of freezing.

One major free radical signal with a g value 2.0036-2.004 is found in skeletal muscle (Davies et al., 1982; Jackson et al., 1985a) and Davies et al. (1982) have obtained evidence that the amplitude of this signal is increased with exhaustive exercise. In our case, excess contractile activity was found to induce both damage to the muscle resulting in a loss of intracellular enzymes and a $70 \pm 20\%$ increase in the amplitude of the major free radical signal. However, the data again do not prove that the increase in free radical concentration was the cause of the damage, or alternatively that it was a consequence of the damage. Another problem which is associated with these studies is that there is no firm relationship established between the esr signal from intact tissues and oxidative damage to tissues. These problems require further study.

IMPLICATIONS OF STUDIES FOR THE TREATMENT OF PATIENTS WITH DUCHENNE MUSCULAR DYSTROPHY AND OTHER NEUROMUSCULAR DISORDERS

The measurements of malonaldehyde in both human biopsy samples and isolated animal muscles have suggested that the process of free radical-mediated lipid peroxidation may be involved in the breakdown of muscles which occurs in muscular dystrophy and in the general process of muscle damage which occurs in a variety of different conditions; this latter hypothesis is supported by studies using ESR techniques (Jackson et al., 1985a). If indeed the mechanisms described which underly muscle damage are relevant to muscular dystrophy, then there are several levels at which pharmacological intervention may be beneficial. These are shown in Table 4.

Table 4. Proposed mechanisms by which muscle damage may be reduced.

(1)   Inhibition of muscle calcium accumulation.

(2)   Inhibition of muscle phospholipases.

(3)   Inhibition of free radical mediated processes.

Muscle calcium has been shown to be elevated in patients with muscular dystrophy (Bodenstein & Engel, 1978; Maunder-sewry et al., 1980; Jackson et al., 1985c) and muscle phospholipase activity is also known to be increased (Tagesson & Henrickson, 1984). Calcium antagonists appear to have some beneficial effects on animal models of muscular dystrophy but have severe side effects and no benefit was seen in patients with Duchenne muscular dystrophy (Emery et al., 1984). An alternative means of reducing muscle calcium accumulation (i.e., by parathyroidectomy) has also been shown to be beneficial in dystrophic hampsters (Palmieri et al., 1980). The mechanism by which this procedure can reduce muscle calcium is unknown but may be related to the claimed pathogenic effect of parathyroid hormone in patients with renal failure (Massry et al., 1983) where tissue calcium levels are known to rise.

Unfortunately, no phospholipase inhibitors specific to muscle are known, but treatment of muscular dystrophy patients with steroids reduces the plasma creatinine kinase activity (Cohen et al., 1972) and steroids are also known to inhibit phospholipase activity in tissues, possibly via interaction with a phospholipase inhibitory protein (Hirata, 1981).

Several groups of workers have examined body fluids of patients with muscular dystrophy for evidence of abnormal production of free radical reaction products or changes in free radical protective mechanisms (Matkovics et al., 1982; Burri et al., 1980; Hunter et al., 1981) and have concluded that there may be some abnormality in this disease. Both plasma and muscle vitamin E levels are apparently normal in patients with muscular dystrophy (e.g., Jackson et al., 1984c) and despite early reports to the contrary (e.g., Bicknell, 1940), it now appears that there is no beneficial effect of short term vitamin E treatment in these diseases (Fitzgerald & McArdle, 1941; Walton & Natrass, 1954).

The situation with selenium metabolism is not so clear. Scandinavian workers claim that both the plasma selenium concentration and whole body turnover of selenium is abnormal in patients with muscular dystrophy (Westermark, 1977; Westermark et al., 1982). We have, however, been unable to demonstrate any abnormality in selenium metabolism in these diseases (Jackson et al., 1985d). Other Scandinavian workers have suggested that oral selenium supplementation may be a beneficial therapy in certain types of muscular dystrophy (Orndahl et al., 1984). We have also undertaken therapeutic trials of vitamin C and zinc, two other agents which may inhibit free radical induced damage (Willson, 1974). Neither agent was found to produce any benefit in patients.

In conclusion, it therefore appears that despite some evidence of increased free radical activity in patients with Duchenne muscular dystrophy, there is little evidence that therapeutic intervention with antioxidants has any beneficial effects on patients. Future work should concentrate on development of better techniques to assess free radical-mediated processes in patients and to determine the effect of therapeutic intervention on these. Analysis of needle biopsy samples of muscle by ESR techniques may be of use in this respect. Another area which requires further study is the applicability of the *in vitro* damage work to the intact animal and man — studies which will problably involve development of *in vivo* models of muscle damage.

Acknowledgements. The author would like to acknowledge the excellent collaboration of Dr. D.A. Jones, Professor M.C.R. Symons and Professor A.T. Diplock. Expert technical assistance was provided by Ms. C. Forte and Ms. L. Burns, and financial support by the Muscular Dystrophy Group of Great Britain and F. Hoffmann-La Roche and Co. Ltd.

REFERENCES

Anand, R., and Emery, A.E.H., 1980, Calcium stimulated enzyme efflux from human skeletal muscle, *Res. Comm. Chem. Path. Pharmacol.*, 28:541.

Anand, R., and Emergy, A.E.H., 1982, Verapamil and calcium-stimulated enzyme efflux from skeletal muscle, *Clin. Chem.*, 28:1482.

Baracos, V., Rodemann, P., Dinarello, C.A., and Goldberg, A.L., 1983, Stimulation of muscle protein degradation and prostaglandin E_2 release by leucocytic pyrogen (Interleukin - 1), *New Eng. J. Med.*, 308:553.

Bicknell, F., 1940, Vitamin E in the treatment of muscular dystrophies and nervous disease, *Lancet*, i:10.

Bodenstein, J.B., and Engel, A.G., 1978, Intracellular calcium accumulation in Duchenne dystrophy and other myopathies: a study of 567,000 muscle fibres in 114 biopsies, *Neurology*, 28:439.

Boveris, A., and Chance, B., 1973, The mitochondrial generation of hydrogen peroxide. General properties and effect of hyperbaric oxygen, *Biochem. J.*, 134:707.

Boveris, A., Cadenas, E., and Stoppani, A.O.M., 1976, Role of ubiquinone in the mitochondrial generation of hydrogen peroxide, *Biochem. J.*, 156:435.

Brady, P.S., Brady, L.J., and Ullrey, D.E., 1979, Selenium, vitamin E and the response to swimming stress in the rat, *J. Nutr.*, 109:1103.

Barri, B.J., Chan, S.G., Berry, A.J., and Yarnell, S.K., 1980, Blood levels of superoxide dismutase and glutathione peroxidase in Duchenne muscular dystrophy, *Clinica Chimica Acta*, 105:249.

Cadenas, E., and Boveris, A., 1980, Enhancement of hydrogen peroxide formation by protophores and iconophores in antimycin-supplemented mitochondria, *Biochem. J.*, 188:31.

Claremont, D., Jackson, M.J., and Jones, D.A., 1984, Accumulation of calcium in experimentally damaged mouse muscles, *J. Physiol.*, 353:57P.

Cohen, L., Morgan, J., and Schulman, S., 1972, Diethyl-stilbestrol: observations on its use in Duchenne's muscular dystrophy (DMD), *Proc. Roy. Soc. Med.*, 140:830.

Commoner, B., Townsend, J., and Pake, G.E., 1954, Free radicals in biological materials, *Nature*, 174:689.

Davies, K.J.A., Quintanilha, A.T., Brooks, G.A., and Packer, L., 1982, Free radicals and tissue damage produced by exercise, *Biochem. Biophys. Res. Comm.*, 107:1198.

Dawson, D.M., 1966, Efflux of enzymes from chicken muscle, *Biochim. Biophys. Acta*, 113:144.

Dillard, C.J., Litor, R.E., Savin, W.M., Dunelin, E.E., and Tappel, A.L., 1978, Effects of exercise, vitamin E and ozone on pulmonary function and lipid peroxidation, *J. Appl. Physiol.*, 45:927.

Duncan, C.J., 1978, Role of intracellular calcium in promoting muscle damage: a strategy for controlling the dystrophic condition, *Experientia*, 34:1531.

Duncan, C.J., Smith, J.L., and Greenaway, H.C., 1979, Failure to protect fog skeletal muscle from ionophore-induced damage by the use of the protease inhibitor leupeptin, *Comp. Biochem. Physiol.*, 63c:205.

Ebashi, S., and Sygita, H., 1979, The role of calcium in physiological and pathological processes of skeletal muscle, in "Current Topics in Nerve and Muscle Research," pp. 73, A.J. Aguayo and G. Karpati, eds., Excerpta Medica, Amsterdam.

Emery, A.E.H., and Skinner, R., 1983, Double-blind controlled trial of a 'calcium blocker' in Duchenne muscular dystrophy, *Cardiomyology*, 2:13.

Farris, M.W., Pascoe, G.A., and Reed, D.J., 1985, Vitamin E reversal of the effect of extracellular calcium on chemically induced toxicity in hepotocytes, *Science*, 227:751.

Fitzgerald, G., and McArdle, B., 1941, Vitamins E and B_6 in the treatment of muscular dystrophy and motor neurone disease, *Brain*, 64:19.

Gee, D.L., and Tappel, A.L., 1981, The effect of exhaustive exercise on expired pentane as a measure of *in vivo* lipid peroxidation in the rat, *Life Sciences*, 28:2425.

Godwin, K.O., Edwardly, J., and Fuss, C.N., 1975, Retention of ^{45}Ca in rats and lambs associated with the onset of nutritional muscular dystrophy.

Guarnieri, C., Ferrari, R., Visoli, O., Caldera, C.M., and Nayler, W.G., 1978, Effect of α-tocopherol on hypoxic-perfused and reoxygenated rabbit heart muscle, *J. Mol. Cell. Cardiol.*, 10:893.

Gutteridge, J.M.C., 1977, The effect of calcium on phospholipid peroxidation, *Biochim. Biophys. Res. Comm.*, 74:529.

Halliwell, B., and Gutteridge, J.M.C., 1984, Lipid peroxidation, oxygen radicals, cell damage and antioxidant therapy, *Lancet*, ii:1396.

Halliwell, B., and Gutteridge, J.M.C., 1985, "Free Radicals in Biology and Medicine," Clarendon Press, Oxford.

Heckley, R.J., 1976, Free radicals in dry biological systems, in "Free Radicals in Biology," 2:135, W.A. Pryor, ed., Academic Press, New York.

Hirata, F., 1981, The regulation of lipomodulin, a phospholipase inhibitory protein, in rabbit neutrophils by phosphonylation, J. Biol. Chem., 256:7730.

Hunter, M.I.S., Brzeski, M.S., and de Vane, P.J., 1981, Superoxide dismutase, glutathione peroxidase and thiobarbitanic acid — reactive compounds in erythrocytes in Duchenne muscular dystrophy, Clinica Chimica Acta, 115:93.

Jackson, M.J., Jones, D.A., and Edwards, R.H.T., 1983a, Vitamin E and skeletal muscle, in "Ciba Foundation Symposium Series No. 101. Biology of Vitamin E," pp. 224, R. Porter and J. Whelan, eds., Pitman, London.

Jackson, M.J., Jones, D.A., and Edwards, R.H.T., 1983b, Lipid peroxidation of skeletal muscle — an in vitro study, Bioscience Reports, 3:609.

Jackson, M.J., Jones, D.A., and Edwards, R.H.T., 1984a, Experimental skeletal muscle damage: the nature of calcium activated degenerative processes, Europ. J. Clin. Invest., 14:369.

Jackson, M.J., Jones, D.A., and Harris, E.J., 1984b, Inhibition of lipid peroxidation in skeletal muscle homogenates by phospholipase A_2 inhibitors, Bioscience Reports, 4:581.

Jackson, M.J., Jones, D.A., and Edwards, R.H.T., 1984c, Techniques for studying free radical damage in muscular dystrophy, Medical Biology, 62:135.

Jackson, M.J., Edwards, R.H.T., and Symons, M.C.R., 1985a, Electron spin resonance studies of intact mammalian skeletal muscle, Biochem. Biophys. Acta, 847:185.

Jackson, M.J., Jones, D.A., and Edwards, R.H.T., 1985b, Vitamin E and muscle diseases, J. Inhert. Met. Dis., 8:Suppl. 1:84.

Jackson, M.J., Jones, D.A., and Edwards, R.H.T., 1985c, Measurements of calcium and other elements in needle biopsy samples of muscle from patients with neuromuscular disorders, Clinica Chimica Acta, 147:215.

Jackson, M.J., Round, J.M., Diplock, A.T., and Edwards, R.H.T., 1985d, Selenium status of patients with Duchenne muscular distrophy, Europ. J. Clin. Invest.,, 15:138.

Jain, S.K., and Shohet, 1981, Calcium potentiates the peroxidation of erythrocyte membrane lipids, Biochem. Biophys. Acts, 642:46.

Jones, D.A., Jackson, M.J., and Edwards, R.H.T., 1983, Release of intracellular enzymes from an isolated mammalian skeletal muscle preparation, Clin. Sci., 65:193.

Jones, D.A., Jackson, M.J., McPhail, G., and Edwards, R.H.T., 1984, Experimental muscle damage: the importance of external calcium, 66:37.

Kar, N.C., and Pearson, C.M., 1979, Catalase, superoxide dismutase, glutathione reductase and thiobarbituric acid — reactive products in normal and dystrophic human muscle, Clin. Chim. Acta, 94:277.

Katz, A.M., 1982, Membrane-derived lipids and the pathogenesis of ischaemic myocardial damage, J. Mol. Cell. Cardio., 14:627.

Massry, S.G., 1982, Parathyroid hormone and the heart, Adv. in Exptl. Med. and Biol., 151:607.

Matkovics, B., Laszlo, A., and Szabo, L., 1982, A comparative study of superoxide dismutase, catalase and lipid peroxidation in red blood cells from muscular dystrophy patients and normal controls, Clinica Chimica Acta, 118:289.

Maunder-Sewny, G.A., Gorodetsky, R., Yaron, R., and Dubowitz, V., 1980, Elemental analysis of skeletal muscle in Duchenne muscular dystrophy using X-ray fluorescence spectrometry, Muscle and Nerve, 3:502.

McCord, J.M., 1985, Oxygen-derived free radicals in postischaemic tissue injury, *New Eng. J. Med.*, 312:159.

Nayler, W.G., Poole-Wilson, P.A., and Williams, A., 1979, Hypoxia and calcium, *J. Mol. Cell. Cardio.*, 11:683.

Orndahl, G., Rindby, A., and Selin, E., 1984, Myotonic dystrophy and selenium, *Acta Med. Scand.*, 211:493.

Palmieri, G.M.A., Nutting, D.F., Bhattacharya, S.K., Bartorini, T.E., and Williams, J.C., Parathyroid ablation in dystrophic hamsters, *J. Clin. Invest.*, 68:646.

Publicover, S.J., Duncan, C.J., and Smith, J.L., 1978, The use of A23187 to demonstrate the role of intracellular calcium in causing ultrastructural damage in mammalian muscle, *J. Neuropath. Exp. Neurol.*, 37:544.

Quintanilha, A.T., and Packer, L., 1983, in "Ciba Foundation Symposium Series No. 101, Biology of Vitamin E", pp. 56, R. Porter and J. Whelan, eds., Pitman, London.

Rodemann, H.P., Waxman, L., and Goldberg, A.L., 1981, The stimulation of protein degradation in muscle by Ca^{2+} is mediated by prostaglandin E_2 and does not require the calcium activated protease, *J. Biol. Chem.*, 257:8716.

Rao, P.S., Cohen, M.V., and Muellar, H.S., 1983, Production of free radicals and lipid peroxides in early experimental myocardial ischaemia, *J. Mol. Cell. Cardio.*, 15:713.

Schanne, F.X., Kane, A.B., Young, A.B., and Farber, J.L., 1979, Calcium dependence of toxic cell death: A final common pathway, *Science*, 206:700.

Siu, G.M., and Draper, H.H., 1982, Metabolism of malonaldehyde *in vivo* and *in vitro* lipids, 17:349.

Smith, M.T., Thor, H., and Orrenius, S., 1981, Toxic injury to isolated hepatocytes is not dependent on extracellular calcium, *Science*, 213:1257.

Snider, M.T., Balke, P.O., Oerter, N.A., Francalonia, N.A., Bull, A.P., Pasko, K.A., and Robbins, M.E., (in press) Lipid peroxidation during muscular exercise in man: inferences from the pulmonary excretion of n-pentane, isopentane and nitrogen, *Proc. Nut. Soc.*.

Soybell, D., Morgan, J., and Cohen, L., 1978, Calcium augmentation of enzyme leakage from mouse skeletal muscle and its possible site of action, *Res. Comm. Chem. Pathol. Pharmacol.*, 20:317.

Suarez-Kurtz, G., and Eastwood, A.B., 1981, Release of sarcoplasmic enzymes from frog skeletal muscle, *Am. J. Physiol.*, 241:C98.

Tagesson, C., and Henriksson, K.G., 1984, Elevated phospholipase A in Duchenne muscle, *Muscle an Nerve*, 7:250.

Walton, J.N., and Nattrass, F.J., 1954, On the classification natural history and treatment of the myopathies, *Brain*, 77:169.

Weglicki, W.B., 1980, Degradation of phospholipids of myocardial membranes, in "Degradative Processes in Heart and Skeletal Muscle," K. Wildenthal, ed., pp. 377, Elsevier/North Holland Biomedical Press.

Westermark, T., 1977, Selenium content of tissues in Finnish infants and adults with various diseases, and studies on the effects of selenium supplementation in neuronal ceroid lipofuscinosis patients, *Acta Pharmacol. et toxicol.*, 41:121.

Westermark, T., Rahola, T., Kallio, A-K., and Suomela, M., 1982, Long-term turnover of selenite-Se in children with motor disorders, *Klin. Padiat.*, 194:301.

Wrogemann, K., and Pena, S.J.G., 1976, Mitochondrial overload: a general mechanism for cell necrosis in muscle diseases, *Lancet*, ii:672.

OXIDATIVE AND PEROXIDATIVE DAMAGE
AND ITS ROLE IN THE AGING PROCESS

David Gershon

Department of Biology
Technion - Israel Institute of Technology
Haifa, Israel

INTRODUCTION

Aging is a phenomenon marked by a progressive decline with time of the capacity of the organism, or given component systems in it, to respond to environmental challenges. This phenomenon can be associated with the accumulation of alterations in structure, composition and function of cells, tissues and organs of multicellular organisms. Aging organisms thus characteristically exhibit a time-associated increase in the propensity to die. Death ensues when one or several activities necessary for the maintenance of viability of cellular or organ systems decline below a critical level. This can happen randomly in any one of an assortment of tissue or organ systems which are essential for the viability of the whole organism (e.g., the brain, heart or lung systems).

The causes of aging have been hypothesized to be either programmed (genetically determined) or stochastic in nature. So far, no convincing evidence has been provided for programmed senescence as a universal phenomenon. Nevertheless, it is widely accepted that genetic determinants involved with development and differentiation exert indirect pleiotropic effects on longevity by affecting post-reproductive stages of the life span. On the other hand, it is quite universally accepted that random detrimental factors play a decisive role in the process of aging.

One of the most important of these random factors is damage caused by free radical reactions which are modulated by genetic and environmental agents. That this is a fundamental factor in aging was first postulated by Denham Harman some thirty years ago. The main premise of Harman's so-called "Free Radical Theory of Aging" is that in aerobic organisms oxygen is the main source of damaging free radical reactions. Because of the high chemical reactivity of active oxygen species and their products, Harman postulated that all cellular components are continually subject to some degree of chemical alterations in a random manner. The following longevity-limiting effects are expected to be triggered by active oxygen species:

a) cumulative oxidative modifications in long-lived molecules such as DNA and the proteins collagen and elastin;

b) modifications in amino acid residues of proteins, particularly cysteine, methionine, tyrosine, tryptophan, and histidine;

c) breakdown of polysaccharides (in the synovial fluids, for instance) through oxidative degradation;

d) changes in the plasma, mitochondrial and lysosomal membranes due to lipid peroxidation;

e) accumulation of metabolically inert components such as lipofuscin ("age-pigment") through oxidative cross-linking and polymerization of polyunsaturated lipids and proteins.

All of the above mentioned effects result in gradual modifications in various cell and tissue functions which reduce organismic adaptability to environmental challenges. Once a certain threshold of alterations is reached, the whole system collapses and death ensues.

The main active forms of oxygen which are encountered by cells throughout life are O_2^-, H_2O_2, OH·, and singlet oxygen. These can be generated by oxygen metabolism in mitochondria, or by detoxification of xenobiotics in the microsomal fraction of the liver. This generation of active species is potentiated by hyperoxia, inflammatory processes and radiation. Lipid peroxides which are by-products of free radical propagation reactions in the cells are also a source of cellular damage. One of the terminal products of lipid peroxidation is malondialdehyde which has been shown to increase in concentration in various tissues of aging organisms. The life span of organisms can also be affected by some active species which are environmental in nature such as aromatic hydrocarbons, NO_2 and O_3.

Defenses against damage caused by active oxygen species have evolved in cells of living organisms whose relative protective capacity as a function of age should be considered. These defenses are:

a) Antioxidants — tocopherols, carotenes, ascorbate and glutathione.

b) Enzyme systems — superoxide dismutases, glutathione peroxidases (GPX), catalase (CAT), and the glutathione (GSH) generating system which consists of glutathione reductase (GR), glucose-6-phosphate (G6PD) dehydrogenase and the GSH synthetic pathway. Also, methionine sulfoxide reductase is probably a protein "repair" enzyme.

c) Turnover of damaged components such as proteins, lipids, and polysaccharides.

d) The DNA repair system.

The latter two types of defense are outside the scope of the present review and will thus not be discussed here in detail. Suffice it to say that both the capacity of old systems to remove and replace damaged components and to repair DNA damage have been shown to decline considerably.

One of the manifestations of the inability of an older system to dispose of altered components has been the accumulation of the ubiquitous aging pigment — lipofuscin. Chio and Tappel (1969) were the first to show that lipofuscin can be formed by the formation of a Schiff's base between the amino groups of denatured proteins, peptides, nucleic acids or phosphatidyl ethanolamine (PE) and malondialdehyde (MDA), or other carbonyls derived from peroxidized polyunsaturated fatty acids (PUFAs). Lipofuscin is therefore made up of a family of related pigment compounds. These pigments, generated during lipid peroxidation, contain a large lipid soluble fraction, probably of peroxidized phospholipid origin. In several cases it has been shown that lipofuscin is formed at a higher rate under elevated temperatures and higher concentrations of oxygen and is correlated with shortened life span. The following scheme is proposed for lipofuscin formation.

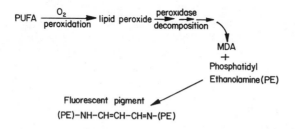

The chromophore thus produced has a maximal fluorescence intensity at 450-470 nm. Often lipofuscin is found in damaged and old tissues to be associated with partially digested mitochondria and incompletely degraded protein molecules which cross-link with peroxidized lipids. This pigment is located in residual bodies which are derived from inactive lysosomes and is thought then to be the result of a defective protein degradation system and reduced breakup of peroxidized lipids characteristic of older cells. Lipofuscin is in many cases a good indicator of tissue damage in old age but does not exert any obvious damaging effect by itself. This, despite the fact that it may occupy large proportions of cell volume in aging animals. For instance, in one-year old rats it occupies 1-2% of the cells in the cerebral cortex and hippocampus and 3% in Purkinje cells. These values increase in two-year old rats to 12%, 10% and 24% in these cell types, respectively (Hirai et al., 1982). Treatment of free living nematodes with the antioxidant α-tocopherol both increased their life span and reduced the rate of accumulation of the pigment in their gut cells (Epstein and Gershon, 1972). Apparently, in tissues of young and healthy animals the rate of anabolic activity is sufficient to dispose of inactive molecules which are associated with the production of this pigment. This is not the case in diseased states or during aging, and thus the accumulation of the pigment under these circumstances.

Substantial age-associated increases in the amount of MDA in a variety of tissues of different animal species were recorded by several groups of investigators. This indicates that oxidized, cellular components accumulate in tissues as a function of age. It is still an open question whether the accumulation of these components in tissues of old organisms is due to higher rates of oxidative damage or due to their inefficient disposal (or, of course, a combination of the two). It is quite certain though that enzymatic protective mechanisms against oxidative damage decline in activity with age.

Studies in our laboratory on the enzymes superoxide dismutase (SOD), glutathione reductase (GR), glutathione peroxidase (GPX) catalase and glucose-6-phosphase dehydrogenase (G6PD), show that they decline considerably in activity in several cell types as a function of age. This is expressed as lowered specific activity per protein in tissue homogenates. Moreover, it is found that defective molecules of some of these enzymes accumulate with age. While the nature of the defects in these enzyme molecules has not been fully studied yet, there is suggestive evidence that oxidative damage to these protein molecules is perhaps an important cause of enzyme inactivation. For instance, it has been shown that glutamine synthetase and pyruvate kinase are inactivated by oxidation of specific histidine residues caused by mixed function oxidases (Fucci et al., 1983). Glyceraldehyde 3-P-dehydrogenase is inactivated by oxidation of a specific cysteine in the active site (Gafni and Noy, 1984). Several proteins and peptides are affected by oxidation of one of their methionine residues. Among these are lysozyme, ribonuclease, α-1-proteinase inhibitor, ACTH, chemotactic factor fMet-Leu-Phe and complement factor C5A (Brot and Weissbach, 1982). There is also substantial accumulation of methionine sulfoxide in old and cataractous lenses (Truscott and Augusteyn, 1977; Garner and Spector, 1980). It is, of course, possible that proteinases and other enzymes involved in the protection of cells against oxidative damage are themselves affected in the same manner as the above mentioned enzymes. If their turnover slows down in cells of old individuals, one would expect a much reduced protection with age. This is what has actually been observed in a variety of systems studied in several laboratories.

CONSEQUENCES OF AGE-ASSOCIATED DAMAGE IN SEVERAL TISSUES

Oxidative damage has been investigated in a variety of tissue types of several organisms and has been implicated as a major factor in aging. A few examples of studies conducted in the author's own laboratory will illustrate this point.

The Erythrocyte

This cell type is an end product of an elaborate series of differentiation steps in the bone marrow. Once this cell emerges into the circulation it has a limited mean life span of about 120 days in man and 60 days in the rat. Due to a series of changes, the cell acquires progressively higher densities; therefore, young, newly emergent erythrocytes can be separated from old ones by density gradient centrifugation. The very old cells (or damaged ones) are removed from the circulation by macrophages in the spleen and liver. The earmarking for this sequestration of old erythrocytes is a peptide antigen ("senescence antigen") which becomes exposed on the membrane surface in old cells. This occurs as a consequence of membrane alterations which allow the antigen to be proteolytically cleaved off of band 3 protein — the anion channel of the cell. The senescence antigen then combines with a physiological auto-antibody, causing opsonization of the cell which leads to its selective phagocytosis by macrophages. It has been shown that the exposure of the "senescence antigen" occurs prematurely in erythrocytes of old rats and humans (Glass et al., 1983; Glass et al., 1985). In parallel, it was shown that the enzympatic activity of G6PD, GPX, GR, and CAT in newly formed red blood cells is reduced in older individuals (Table 1). This is in addition to the decline which occurs during the aging of the individual cells. Concomitant with this, it was found that increased levels of products of lipid peroxidation such as malondialdehyde accumulated in the cells of older individuals (Table 2). Preliminary experiments indicate that young erythrocytes subjected to superoxide, generated externally by xanthine oxidase and acetaldehyde, show premature exposure of the senescence antigen. These findings suggest that peroxidative damage to the cell membranes may be an important determining factor of its longevity in the circulation. Lipid peroxidation, and perhaps the oxidation of protein components, lead to membrane alterations which facilitate proteolytic activity that results in the formation of the senescence antigen. This results in an early sequestration of the erythrocyte in old individuals. Since there are no distinct changes in the hematocrit of healthy older individuals, these findings indicate that erythrocyte turnover in old individuals is more rapid than in young indivduals. In fact, the same conclusion has been reached from half-life studies of erythrocytes in rats of various ages (Glass et al., 1983). The increased demand for the production of erythrocytes in senescent organisms can be considered as a severe stress to the hematopoietic system. This stress situation may manifest itself in old individuals in a retarded response of this system to severe bleeding and a slower replenishment of erythrocytes in the circulation.

Table 1. Enzyme activities in age fractionated erythrocytes from young and old rats.

Cell Fraction*	G6PD		GPX		GR		CAT	
	Young	Old	Young	Old	Young	Old	Young	Old
I	25.3 ±1.4	19.0 ±0.8	0.484 ±0.028	0.287 ±0.029	7.78 ±0.30	4.09 ±0.30	5.25 ±0.35	2.94 ±0.12
II	19.8 ±1.2	15.3 ±1.3	0.424 ±0.022	0.236 ±0.014	5.22 ±0.59	3.18 ±0.35	3.63 ±0.14	2.61 ±0.10
III	15.5 ±1.5	13.0 ±0.8	0.258 ±0.011	0.150 ±0.020	3.10 ±0.52	2.49 ±0.33	2.76 ±0.12	2.31 ±0.13
IV	11.9 ±1.0	10.5 ±0.7	0.207 ±0.022	0.078 ±0.012	1.78 ±0.50	1.00 ±0.20	2.44 ±0.16	1.83 ±0.13

G6PD   - activity in units per 10^6 cells.

GPX   - activity in units per 10^6 cells.

GR   - activity in units per 10^9 cells.

CAT   - catalase activity in units per 10^9 cells.

*Fractions I-IV increasing density and age of cells.

Table 2. Erythrocyte membrane lipid peroxidation as a function of cell and donor age (rats).

	Amounts of malonaldehyde (nmol/10^6 cells)	Chromolipid fluorescence (units)
Fraction I (young cells) of young animals	10.8 ± 3.2	19.3 ± 3.0
Fraction IV (old cells) of young animals	32.5 ± 3.2	44.0 ± 3.0
Fraction I (young cells) of old animals	37.8 ± 1.3	44.2 ± 3.0
Fraction IV (old cells) of old animals	48.6 ± 2.6	57.8 ± 4.0
Erythrocytes incubated with xanthine/xanthine oxidase for 2 h	21.8 ± 1.3	20.7 ± 0.6
Erythrocytes incubated with xanthine/xanthine oxidase for 4 h	48.7 ± 5.1	55.7 ± 4.0

Thiobarbituric acid measurements are given in nmol of malonaldehyde per 10^6 cells reacting with thiobarbituric acid. Fluorescence measurements were done by the procedure of Jain & Hochstein (1980) and are given in arbitrary units of fluorescence per 10^6 cells, at 390 nm after excitation at 460 nm.

The Ocular Lens

The lens is a peculiar, highly differentiated organ whose structure and functions are geared to focusing light on the retina and to allow maximal transmittance of this light. The optical properties of the lens, which are controlled by a series of specialized proteins — the crystallins, are preserved for very long periods of time. This despite the fact that the lens is almost completely devoid of the capacity to synthesize and degrade cellular components such as proteins. Opacities develop in the lens when proteins and other components are damaged in the cells, eventually resulting in cataracts and blindness. Cataracts are highly prevalent among individuals of advanced ages. In cataractous lenses (particularly "senile" cataracts), it was found that 90% of the thiol groups of cysteins are oxidized, and that 45-60% of the methionine appears as methionine sulfoxide (Truscott and Augusteyn, 1977; Garner and Spector, 1980). The activity of enzymes which are involved in antioxidative protection such as SOD and G6PD declines considerably with age (Table 3). Moreover, inactive forms of the enzyme SOD have been detected in lenses of senescent rats as revealed by immune titration studies (Dovrat and Gershon, 1981).

Table 3. Activity in rat lenses as a function of age.

AGE	SOD	GGPD
2 days	0.950 ± 0.09	74 ± 10
6 months	0.347 ± 0.08	10 ± 2
27 months	0.237 ± 0.08	5 ± 1

SOD (units/mg protein).
GGPD (μunits/mg protein).

The understanding of how the lens maintains its optical activity for many years without replacing damaged components may provide essential clues as to the underlying mechanisms leading to aging of other systems. This could lead to devising means which will decelerate the process of aging. We would like to suggest that some of these clues are the following: a) the lens is exposed to very low concentrations of oxygen due to the fact that it is avascular and does not get any direct blood supply; b) with the exception of a unicellular layer of epithelium, the lens cells lack mitochondria, and thus do not derive their energy from the electron transport system which generates superoxide; c) the lens has a very high concentration of glutathione relative to other tissues; d) many diurnal animals species have high concentrations of ascorbate in the aqueous humor which bathes the anterior part of the lens. These observations suggest that despite the constant exposure to light, oxidative damage accumulates very slowly in the lens, and that it is only in senescent organisms that the optic functions are likely to be seriously affected.

The Nematode System-Use of Antioxidants to Increase the Life Span

Free-living nematodes can be raised in culture axenically and provide age-synchronized populations. Under the culture conditions used, *Caenorhabditis leriggsae* shows a mean life span of about 30 days. Electronmicroscopic studies have shown that lipofuscin accumulates in very large amounts in the gut epithelium cells of this organism as it ages (Epstein and Gershon, 1972). Concomitantly, extensive damage occurred in the plasma and endoplasmic reticulum membranes of these cells. This membrane damage was suggested to be the result of lipid peroxidation (Epstein and Gershon, 1972). It was, therefore, postulated that increased amounts of antioxidants in the nutrient medium might reduce the age-associated membrane damage and also the degree of accumulation of lipofuscin in gut epithelium cells. It was also thought that it would be of fundamental interest to observe whether reduction in peroxidative damage of cellular components of the aging organism might extend its life span. As can be seen in Figure 1, the mean and absolute life spans of *C.briggsae* were extended approximately 30% by the addition of α-tocopherol to the nutrient medium. There was also a considerable delay in the accumulation of lipofuscin in the treated animals. A striking finding was that the vitamin E was most effective in extending the life span when given in the first third of its life span. It had only a marginal effect on the life span if added after the age

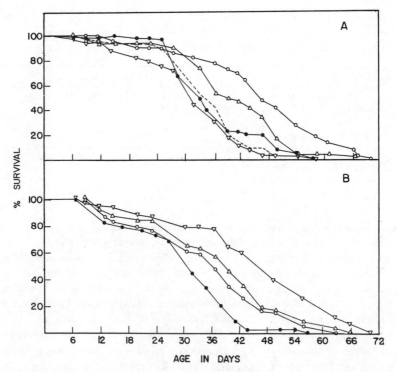

Figure 1. Effect of α-tocopherol (400 μ g/ml) on the life span of *Caenorhabditis briggsae*.
A) Survival curves of animals supplied with the vitamin in the first day (o), 10 (Δ), 20 (●), and 30 (∇) days of age. (---) represents an untreated control population. B) α-tocopherol added to newly hatched animals and removed at the ages of 6 (o) and 10 (Δ) days. (●) and (∇) are the survival curves of *C.briggsae* grown with no vitamin in the medium and animals continuously exposed to the vitamin, respectively.

of 10-15 days. Since this nematode reaches reproductive maturation at 5-7 days of age and completes its growth by day 10, these results suggest that peroxidative damage in the early stages of development and maturation may play a decisive role in the determination of the life span. It would be of utmost importance to determine what are the crucial cellular and tissue components which are most affected by oxidative damage during development and whether they exert a delayed effect on post reproductive stages of the organism.

The effect of antioxidants on the life span and on the function of various tissues in old organisms has been investigated in various other organisms. The results have been equivocal. Further standardized experiments should be carried out in order to decide how important a role does oxidative damage play in age-related deterioration of functions of various physiological and biochemical systems in the organism. The results so far seem to merit extensive experimentation with various anti-oxidants. The uptake by and retention in cells of these antioxidants must be verified, an experimental feature which is missing in the studies conducted to date.

SUMMARY

In summary, the work presented above indicates the following: The continual effect of active oxygen species, whether internally generated or from external sources, on cellular and extracellular components probably limits the viability of cells and whole organisms. This is most probably due to an increasing inability of older systems to protect against such active species and to replace damaged components at the rate required for maintenance of viability. The reduction in the protective ability of the aging organism originates from damage to the enzymes which are directly involved in scavenging of superoxide and hydrogen peroxide and those involved in the maintenance of glutathione and $NADPH_2$ levels in the cells. Further experimental work is required to completely establish these suggestions.

REFERENCES

Brot, N., and Weissbach, H., April 1982, *Trens Bioch. Sci.*, 137-139.

Chio, K.S., and Tappel, A.L., 1969, *Biochemistry*, 8:2821.

Dovrat, A., and Gershon, D., 1981, *Exp. Eye Res.*, 33:651.

Epstein, J., and Gershon, D., 1972, *Mech. Age. Dev.*, 1:257.

Fucci, L., Oliver, C.N., Boon, M.J., and Stadtman, E.R., 1983, *Proc. Natl. Acad. Sci. USA*, 80:1521.

Gafni, A., and Noy, N., 1984, *Mol. Cell. Biochem.*, 59:113.

Garner, M.H., and Spector, A., 1980, *Proc. Natl. Acad. Sci. USA*, 77:1274.

Glass, G.A., Gershon, D., and Gershon, H., 1985, *Exp. Hematol.*.

Glass, G.A., Gershon, H., and Gershon, D., 1983, *Exp. Hematol.*, 11:987.

Hirai, S., Okamoto, K., and Morimatsu, M., 1982, in "Lipid Peroxides in Biology and Medicine," pp. 305-315, Academic Press, Inc.

Truscott, R.J.W., and Augusteyn, R.C., 1977, *Bioch. Biophys. Acta*, 492:43.

RECOMMENDED READING

Freeman, B.A., and Crapo, J.D., 1982, Biology of disease: free radicals and tissue injury, *Lab. Invest.*, 47:412.

Harman, D., 1981, The aging process, *Proc. Natl. Acad. Sci. USA*, 78:7124.

Lippman, R.D., 1983, Lipid peroxidation and metabolism in aging: a biological, chemical and medical approach, *Review of Biological Research in Aging*, 1:315, Alan Liss Inc.

Sohal, R.S., ed., "Age Pigments", 1981, Elsevier/North Holland.

Wolff, S.P., Garner, A., and Dean, R.T., January 1986, Free radicals, lipids and protein degradation, *Trends Bioch. Sci.*, 27-31.

INDEX

effect on glutathione, 191
oxidative effects, 187
tissue damage, 187, 188
Exhaustion, 188
Exhaustive swimming, 200
Exposed organs, 122
Exposure
to light, 97
to oxygen, 97
Exudative diathesis, 17
FAD, 148
semiquinone, 150
Fast flow, 95
Fatigue, 188
Fatty acid autoxidation
diene, 70
tetraene, 76
triene, 76
Fatty acids, 9, 55, 60, 62, 199
Fenton reaction, 41, 145
Ferredoxin, 104
Ferricyanide, 151
Ferrous ions, 46
Fish meal, 20
Flash photolysis, 40, 82
Flavins, 107
Flavoprotein, 2
Fluorescence, 118
Fluorimetric method, 17
Fluorimetry, 30
Focal necrosis, inflammation, 188
Formate, 45
Forster criteria, 118
"Free radical clock", 77
Free radical
detection, 170
inhibitors, 69
kinetics, 75
metabolism, 95
reactions *in vitro*, 83
reactions *in vivo*, 83
scavenger, 11
Free radicals in muscle tissue
detection, 201
identification, 201
Furan rings, 92
Furocoumarines, 133

Gamma-rays, 37
g-factor, 87, 89, 205
Glucose-6-phosphate dehydrogenase, 214
Glucose-6-phosphatase, 10
Glutamine synthetase, 214
Glutathione, 7, 10, 18, 82, 83, 95, 104, 124, 146
Glutathione disulfide, 18, 104
Glutathione effects, 192
in liver, 191
in muscle, 191
in plasma, 191
intracellular levels, 190
Glutathione peroxidase, 15, 19, 25, 104, 146, 171, 177, 211, 214
amino acid sequence, 32
assay, 27, 28
biometric catalysts, 32
catalytic cycle, 28
crystallographic data, 32
ethnic variation, 31
first isolation, 32
in blood, 20, 24
molecular weight, 32
properties, 27
reaction mechanism
structure, 32
synthesis, 32
turnover, 194
Glutathione reductase, 27, 29, 146, 214
Glutathione S-transferase, 29
Glyceraldehyde 3-dehydrogenase, 214
Glycolipids, 60
Glycopeptide antibiotics, 175
Granulocytes, 143
Graphite furnace, 16
Ground state triplet, 55
GSH, see glutathione
Guanine, 107
Gyromagnetic ratio, 86, 87
H atom abstraction, 76
H atom donors, 75
Haber-Weiss reaction, 145
Half-lives, 5, 9
Halogen ions, 145
Harmful effects
of light, 101

of oxygen, 101
protection against, 101
2nd harmonic, 101
Heart, 211
disease, 55
Hemin-acridines, 184
Hemoglobin, 18
Hemoprotein, 127
Hepatosis dietetica, 17
Herbicides, 157
toxicity, 162
Hetero-atom, 41
Heterocyclic atom, 7
Heterocyst, 105
Heterolytic scission, 128
High energy particles, 37
High pressure liquid chromatography, 10, 73, 113
2D-high-field NMR, 173
Hindered phenols, 58
Hippocampus, 213
Histidine, 107, 124, 212
Homolytic decomposition, 60
Homolytic scission, 57, 128
Hormone, 9
Horseradish peroxidase, 132
HPLC, see high pressure liquid chromatography
Human health, 16, 21
Human nutrition, 21
Human nutritionists, 26
Hydrated electron, 38, 40
optical absorption, 39
Hydrazines, 2
Hydrazone derivatives, 10
Hydride, 17
Hydrogen abstraction, 82
Hydrogen atom, 38, 44
abstraction, 42, 44, 50, 57, 58
abstraction rates, 63
transfer rate constants, 75
Hydrogen peroxide, 19, 27, 56
Hydrogen transfer, 82
Hydrolytic enzymes, 143
Hydroperoxides, 9, 11, 18, 47, 48, 109
Hydroperoxyeicosatetraenoic acids, 63
Hydroquinones, 50, 104

Hydroxy fatty acids, 10
Hydroxy peroxyl radical, 49
Hydroxyacids, 25
4-hydroxy-Alkenals, 10
4-hydroxy-alkenals, 9
Hydroxycyclohexadienyl radical, 39
6-hydroxydopamine, 129
Hydroxyl radical, 3, 38, 40, 42, 56, 83
Hydroxyl-adducts, 42
Hydroxylated radicals, 42
Hyperfine interaction, 92
Hyperfine pattern, 88
Hypericin, 107
Hyperoxia, 212
Hypochlorite, 111, 146
Hypochlorous acid, 111
Hypohalous acid, 132
Hypoxia, 198
injury, 200
Imidazole, 151
Immune complexes, 152
Immune titration studies, 216
Immunoglobulin G, 145
Impaired fat absorption, 26
Induced transition, 87
Inflammation, 1, 66, 160
Inflammatory conditions, 25
Inflammatory processes, 212
Inositol tris phosphate, 152
Intercalative binding, 178
Internal viscosity, 101
Intracellular oxygen, 98, 101
Iodine, 17
Ionizations, 37
Ionizing radiation, 4, 81
Ionophore, 20
Iron-containing SOD, 103
Irradiated water, 38
Irritation, 160
Isocyanides, 151
Isomerization, 73, 92, 94
Isopentane excretion, 205
Joint inflammation, 153
Keshan disease, 22
Kesterson National Wildlife Refuge, 25
Ketones, 10, 48, 107, 109
"Krypto"-oxy radicals, 51

DATE DUE

APR 2 6 1991		
JUN 3 0 1992		